ADVANCES IN PHOTODETECTORS - RESEARCH AND APPLICATIONS

Edited by **Kuan Chee**

Advances in Photodetectors - Research and Applications
http://dx.doi.org/10.5772/intechopen.74896
Edited by Kuan Chee

Contributors

Richard Ocaya, Raffaello Papadakis, Hu Li, Ioanna Deligkiozi, Caiming Sun, Vera Gradišnik, Darko Gumbarević, Hai-Zhi Song, Yessenia Jauregui-Sánchez, Pere Clemente, Pedro Latorre-Carmona, Jesus Lancis, Enrique Tajahuerce, Yijun Zhang, Gangcheng Jiao, Kuan W. A. Chee

First published in London, United Kingdom, 2019 by IntechOpen
IntechOpen is the global imprint of INTECHOPEN LIMITED, registered in England and Wales, registration number: 11086078, The Shard, 25th floor, 32 London Bridge Street
London, SE19SG – United Kingdom
Printed in Croatia

British Library Cataloguing-in-Publication Data
A catalogue record for this book is available from the British Library

Additional hard and PDF copies can be obtained from orders@intechopen.com

Advances in Photodetectors - Research and Applications, Edited by Kuan Chee
p. cm.
Print ISBN 978-1-78985-621-7
Online ISBN 978-1-78985-622-4
eBook (PDF) ISBN 978-1-78985-389-6

For EU product safety concerns:
IN TECH d.o.o., Prolaz Marije Krucifikse Kozulić 3, 51000 Rijeka, Croatia,
info@intechopen.com or visit our website at intechopen.com.

Meet the editor

Kuan W. A. Chee is Outstanding Professor at the Shandong Academy of Sciences and head of the Optoelectronics Materials Laboratory. He has a diverse background in leading high-tech industries and academia. He has published in several top journals, books, and conference proceedings and has given numerous talks at international conferences and seminars in science and engineering. His research covers electronic design, nano- and post-CMOS electronics, and photonic and optoelectronic devices. He has received prizes and awards such as from the Institution of Engineering and Technology, UK; the Institute of Physics, UK; the Institute of Electrical and Electronics Engineers, USA; the Royal Academy of Engineering, UK; the Materials Research Society; and the Cambridge Philosophical Society. Prof. Chee is supported by a science and engineering talent program in China.

Contents

Preface

This edited volume is a collection of reviewed and relevant research chapters covering state-of-the-art developments in research and technology within the field of photodetectors.

Contributions from scholars and experts are pertinent to photodetector device physics as well as technology and applications.

Followed by an introduction in the first section, section 2, "Materials and Devices," includes chapters on topics including: photoconductive interlocked molecules and macromolecules; ion-beam modified terahertz GaAs photoconductive antenna; energy bandgap engineering of transmission-mode AlGaAs/GaAs photocathode; and a-Si:H p-i-n photodiode as a biosensor.

Section 3, "Technology and Applications," includes chapters on topics including: single-pixel imaging using photodiodes; overcoming the bandwidth-quantum efficiency tradeoff; and linear CCD based spectrometry using a FPGA-centered design methodology. This work will appeal to students and professionals interested in studying and dealing with photodetector design and technology, as well as the wide gamut of related applications.

Kuan Chee
University of Nottingham , Ningbo, People's Republic of China
Laser Research Institute, Shandong Academy of Sciences, People's Republic of China
Faculty of Electrical Engineering and Computer Science, Ningbo University,
Ningbo, People's Republic of China

Section 1

Introduction

Chapter 1

Introductory Chapter: Photodetectors

Kuan W.A. Chee

Additional information is available at the end of the chapter

http://dx.doi.org/10.5772/intechopen.82045

1. Introduction

Modern day electronic communications, industrial electronics, analytical equipment, medicine and healthcare, automotive and transport, etc. widely employ photodetectors, also known as photosensors, primarily as optical receivers to convert light into electrical signals. These devices may receive the transmitted optical pulses, or sense light or other electromagnetic radiation. Nevertheless, the photodetectors may be classified according to their light detection mechanisms, *viz.* the photoemission or photoelectric effect, thermal effect, polarisation effect, photochemical effect, or weak interaction effects. Photodetectors that employ semiconductors operate on the principle of electron-hole pair creation upon light irradiation. When a semiconductor material is illuminated by photons having energies greater than or equal to its bandgap, the absorbed photons promote valence band electrons into the conduction band, thus leaving behind positively charged holes in the valence band. Conduction band electrons (valence band holes) behave as free electrons (holes) that can diffuse in a concentration gradient, or drift under the influence of an intrinsic, or externally applied, electric field. The photogenerated electron-hole pairs due to optical absorption may recombine and re-emit light, unless subjected to an electric field-mediated separation to give rise to a photocurrent, which is a fraction of the photogenerated free charge carriers collected at the electrodes of the photodetector structure. The magnitude of this photocurrent at a given wavelength is directly proportional to the incident light intensity.

In this chapter, we introduce some representative photodetectors, their properties, performance and applications, as applied in the various design configurations. We also address sensing and detection in the electromagnetic spectrum spanning from the ultraviolet and visible, to infrared and terahertz.

2. Photodetection mechanisms

Heinrich Hertz discovered in 1887 that ultraviolet light illumination of electrodes generates electric sparks more easily. While studying black-body radiation in 1900, Max Planck suggested that energy carried by electromagnetic waves could only be quantised into units of discrete packets known as photons or quanta. Albert Einstein advanced the foregoing light energy packet hypothesis to explain experimental results using the notion of the photoelectric effect. The light beam photons have a characteristic energy proportional to the frequency of the light. When the light beam irradiates a material, the energy of the photon, if sufficiently high, is absorbed to liberate the electron from atomic bonding, and the remaining photon energy contributes to the free electron's kinetic energy. For photon energies too low to be absorbed, they are re-emitted. However, if the electron acquires energy surpassing the work function of the material, it is ejected as a photoelectron. Whilst the maximum kinetic energy of the emitted photoelectron depends on the frequency of the irradiance, the photoelectron ejection rate (or magnitude of the photoelectric current) is directly proportional to the intensity of the incident light.

Other than microchannel plate detectors, a range of photodetectors operate on the basis of the photoelectric or photoemission effect. Gaseous ionisation detectors detect photons having sufficient energy to ionise gas atoms or molecules, and the current flow due to the electrons and ions generated by the ionisation can be measured. Photomultiplier tubes or phototubes contain photocathodes that emit electrons when illuminated, thus conducting a current proportional to the light intensity. The thermal effect is realised when the incident photons cause electrons to transition into the mid-gap states, which then relax into the lower bands, thus leading to phonon generation and heat dissipation. The rise in temperature in turn modifies the electrical properties of the device (e.g., thermopile, pyroelectric detector, cryogenic detector, bolometer, etc.) material, such as its electrical conductivity. The polarisation effect is so called when the incident photons alter the polarisation states of appropriate materials, thereby modulating the refractive index (i.e., photorefractive effect); this is exploited in holographic data storage. Photochemical effects in photodetectors occur when chemical changes in the material are induced by the incident photons. Examples include photoreceptor cells in the retina, or photographic plates. Finally, weak interaction effects occur when secondary effects are induced by photons, such as in photon drag detectors or gas pressure changes in opto-acoustic detectors (e.g., Golay cells).

3. Types of photodetectors

Photodetectors may be configured in unique ways for various applications. For example, single sensors may detect overall light intensities. A 1-D array of photodetectors may be used to measure the distribution of light along a line, such as in a spectrophotometer or a line scanner. Moreover, a 2-D array of photodetectors may be used to derive images from the light intensity profile, when applied as an image sensor. Focal-plane arrays (FPAs) are devices consisting of an array of light-sensing pixels or active pixel sensors (APS) at the focal plane of a lens, and are most commonly adopted for imaging (photos or videos) or non-imaging (spectrometry, LIDAR and wave-front sensing) purposes. In radio telescopes, the FPA usually refers to 2-D devices that are sensitive in the infrared. Other image sensors, such as charge-coupled device (CCD) or CMOS sensors, operate in the visible regime. An anti-reflective coating or a surface-plasmon antenna is sometimes used on a photodetector, to enhance the optical absorption or photogeneration of charge carriers (or photocurrent response), respectively. By embedding an ultrathin semiconductor absorption layer into a Fabry-Pérot resonant cavity, resonant cavity enhanced photodetectors can be realised, to boost the quantum efficiency or bandwidth-efficiency product, and provide superior wavelength selectivity and high speed photoresponse for wavelength division multiplexing (WDM) systems. Subwavelength microcell gratings affixed in close proximity to the optical absorber can enable near-field enhancement of optical absorption through strong electromagnetic field confinement [1].

Photovoltaic photodetectors resort to the internal electric field of a *p-n* or Schottky junction to achieve the charge separation and photocurrent generation. Solar cells are similar to photovoltaic photodetectors, which also absorb light and convert it into electrical energy, through the photovoltaic effect. The *p-n* junction photodetectors include designs consisting of a simple *p-n* junction, or *p-i-n* photodetectors incorporating a nominally undoped semiconductor layer between the *p-* and *n*-regions, or phototransistors combining a photodiode and an additional *n*-region. At equilibrium, the presence of the ionised acceptors and donors within the space charge region (SCR) sets up an internal electric field at the junction. Therefore, electron-hole pairs generated inside the SCR, or within the minority carrier diffusion length from the edges of the SCR, will be separated by the built-in electric field and contribute to the photocurrent. The width of the SCR is inversely related to the dopant concentration in the material, but its expansion may be modulated by reverse biasing, which concomitantly increases the internal electric field at the junction so as to enhance the efficiency of electron-hole pair separation. To improve the photoresponse speed, the electrical resistivity of the photodetector material may be reduced through increasing the dopant concentration, but a nominally undoped layer of a thickness largely determining the SCR width may be introduced between the *p-* and *n*-regions to form the *p-i-n* structure. With a lower resistivity and a wider SCR width (and hence lower capacitance), the *p-i-n* structure is well suited for high-speed IC applications.

Avalanche photodiodes are designed with high *p-* and *n*-type doping to intensify the junction electric field. With a reverse bias sufficiently high (100–400 V) such that the internal electric field approximates the critical breakdown field, the acceleration of the photogenerated charge carriers within the SCR is able to ionise the lattice atoms, hence resulting in an avalanche multiplication of charge carriers. The corresponding gain is typically of the order of 10–20 in these cases. Avalanche photodiodes are well suited for fibre optic systems requiring low optical power levels with quantum efficiencies eclipsing 100%.

Phototransistors are similar to photodiodes, except that an additional *n*-region is included in the photodetector design. The phototransistor comprises a photodiode with an internal gain, and it can be represented as a bipolar junction transistor enclosed in a transparent case through which photons are allowed to irradiate the base-collector junction. The electrons generated by the absorbed photons in the base-collector junction SCR are injected into the base, and the photocurrent is amplified. Nevertheless, while a phototransistor is generally a few orders of magnitude more sensitive than the photodiode, the photoresponse speed is much slower. Polysilicon- [2], zinc oxide- [3], or organic polymer-based [4] thin film transistors (TFTs) have been adopted as photodetectors for optical interconnects, ultraviolet imaging and large area displays/flexible substrates, respectively.

Schottky junction photodetectors include Schottky barrier photodiodes and metal-semiconductor-metal (MSM) photodiodes. In the former, the Schottky junction is formed between a metal and a doped semiconductor. Analogous to that formed at the *p-n* junction, the SCR is comparable, and its width can be modulated in tandem with the built-in electric field proportional to the reverse bias to the Schottky junction photodetector. Typically, an ultrathin, semi-transparent metal layer, for example, Au of about 10-nm thick, is used as the Schottky contact, which allows transmissivity up to 95% and around 30% for infrared and ultraviolet, respectively. MSM photodiodes are designed with two Schottky contacts, with one Schottky junction reversed-biased to support an elongated SCR width, and the other, forward biased. Typically, the semiconductor material is nominally undoped, and hence, the SCRs are spatially extended into the device. The reversed-biased Schottky junction generates the photocurrent, whereas the forward-biased Schottky junction acts as a highly efficient charge carrier collector.

In photoconductors, an electric field is applied across a layer of a semiconductor through electrically biased ohmic contacts on either side, leading to the collection of charge carriers. Photoresistors, light-dependent resistors (LDRs) or photoconductive cells change electrical resistivity according to the light intensity, hence exhibiting photoconductivity. Such devices have a higher gain, as the response of photoconductors is typically several orders of magnitude greater than that of the photovoltaic detector counterpart, based on a given material. However, for photoconductors, the bandwidth, infrared sensitivity, ultraviolet-visible contrast and a range of other key performance parameters are inferior to that of other types of photodetectors. Hence, the scope of potential applications is significantly limited.

Rewritable nanoscale photodetectors have been demonstrated based on insulating oxide ($LaAlO_3/SrTiO_3$) interfaces [5], exhibiting electric field-tunable photoconductive response within the electromagnetic spectrum ranging from the visible to near-infrared. The integration of nanoscale photodetectors based on nanodots and nanowires has also benefited from recent innovations in subwavelength imaging beyond the diffraction limit, by adapting (plasmonic) metamaterials for superlenses suitable for superresolution, near-field light focussing [6].

4. Performance figures of merit

High sensitivity at the operating wavelength, short response times, linear response over a wide range of light intensities, minimum noise contribution, stability of performance characteristics, reliability, low bias voltage and low cost are amongst the photodetector requirements for technology adoption. State-of-the-art graphene-on-diamond photodetectors have been demonstrated to exhibit superior responsivity and photocurrent, as well as open circuit voltage [7]. Particularly, in high-speed optical data communications, photodetectors must also be highly responsive to photoexcitation, yet immediately/rapidly relax to the ground state after the light source is switched off. However, the excited non-equilibrium state is usually maintained for a finite amount of time through an effect known as persistent photoconductivity, owing to long recombination times that originate from charge carrier trapping by bulk defects (vacancies or impurities) and surface states. The photodetector may be characterised by various figures of merit such as the spectral response, quantum efficiency, responsivity, bandwidth, gain, noise equivalent power (NEP), dark current, response time and detectivity. The spectral response characterises the photodetector response with respect to the photon frequency. The quantum efficiency is the measure of the number of charge carriers generated per photon. The responsivity is the ratio of the output electrical current to the input optical power to the photodetector. The NEP is the minimum amount of optical power required to generate a signal in the presence of noise in the photodetector. The specific detectivity is the reciprocal of NEP normalised to the square root of the photodetector active area-bandwidth product. The gain is the ratio of the output electrical current to the photogenerated current directly generated by the incident photons. The dark current is a measure of charge carrier flow through a photodetector in the absence of an optical input. The response time is the time needed for a photodetector to rise from 10 to 90% of the final output. The noise spectrum is the intrinsic noise voltage/current as a function of frequency, which can be represented as a noise spectral density. The RF output is constrained by the nonlinearity of the photodetector. All in all, having a large angular acceptance, high temporal resolution, as well as high spectral and energy resolution, may also be crucial design considerations for a high-performance photodetector.

5. Conclusion

For a comparison of the viability and performance of photodetectors, an in-depth understanding of their figures of merit is essential. The insights underpinning the physics and technology of various photodetector designs and configurations must be conscientiously examined for successful implementation and integration of high-performance photodetection and optoelectronic sensing within the relevant wavelength ranges, on low-cost substrates or CMOS-compatible substrates. New device concepts and techniques to develop monolithic integration of optoelectronic materials on a single substrate may permit revolutionary ultrafast and ultrasensitive near-field photodetection at high spatial, temporal and spectral resolution.

Acknowledgements

The following grants are acknowledged: project number 61650110517 supported by the National Natural Science Foundation of China and project numbers 2014A610154 and 2017A610095 supported by the Natural Science Foundation of Ningbo.

Author details

Kuan W.A. Chee

Address all correspondence to: kuan.chee@nottingham.edu.cn

Department of Electrical and Electronic Engineering, Faculty of Science and Engineering, University of Nottingham Ningbo China, Ningbo, People's Republic of China

References

[1] Zohar M, Auslender M, Hava S. Ultrathin high efficiency photodetectors based on subwavelength grating and near-field enhanced absorption. Nanoscale. 2015;**7**(12):5476-5479

[2] Rangarajan B, Brunets I, Holleman J, Kovalgin AY, Schmitz J. TFTs as photodetectors for optical interconnects. In: Proceedings of the 12th Annual Workshop on Semiconductor Advances for Future Electronics and Sensors. Utrecht, The Netherlands: Technology Foundation (STW); 2009. pp. 52-54

[3] Ku CJ, Reyes P, Duan Z, Hong W-C, Li R, Lu Y. $Mg_xZn_{1-x}O$ thin-film transistor-based UV photodetector with enhanced photoresponse. Journal of Electronic Materials. 2015; **44**(10):3471-3476

[4] Hamilton MC, Kanicki J. Organic polymer thin-film transistor photosensors. IEEE Journal of Selected Topics in Quantum Electronics. 2004;**10**(4):840-848

[5] Irvin P, Ma Y, Bogorin DF, Cen C, Bark CW, Folkman CM, et al. Rewritable nanoscale oxide photodetector. Nature Photonics. 2010;**4**:849-852

[6] Memarian M, Eleftheriades GV. Light concentration using heterojunctions of anisotropic low permittivity metamaterials. Light: Science & Applications. 2013;**2**:e114

[7] Yuan Q, Lin C-T, Chee KWA. Carbon sp^2-on-sp^3 technology: A guide for future carbon based devices. In: Zhu J, Jin A, Zhu D, editors. New Trends in Nanotechnology, Material and Environmental Science. Saarbrücken, Germany: AkademikerVerlag; 2018. ISBN-13: 978-620-2-21118-5, ISBN-10: 6202211180

Section 2

Materials and Devices

Chapter 2

Photoconductive Interlocked Molecules and Macromolecules

Raffaello Papadakis, Ioanna Deligkiozi and Hu Li

Additional information is available at the end of the chapter

http://dx.doi.org/10.5772/intechopen.79798

Abstract

Organic compounds and materials with photoconductive properties have been studied for many years because of their importance in many technological applications such as dye-sensitized solar cells, photodiodes, photoresistors, electronics, biomolecular sensing, etc. For multiple purposes, such molecules require intense protection from various factors which can decrease their durability and cause fatigue. Interlocked molecules and macromolecules involving photoconductive organic components and various types of macrocycles, such as cyclodextrins, cyclophanes, or macrocyclic ethers, are promising candidates for new photoconductivity-related applications. In this chapter, a review in this emerging research area in materials science and technology is provided. Focus is placed on photoconductive (poly)rotaxanes and (poly)catenanes. Various types of such materials and compounds are reviewed, and recent examples are provided. The relation between their structure and photoconductive behavior is discussed.

Keywords: photoconductivity, interlocked molecules, rotaxanes, catenanes, photocurrent responses, photosensitivity

1. Introduction

In recent years, a new class of supramolecular assemblies has gained the attention of the scientific community [1]. Supramolecular chemistry is a rapidly increasing research field which focuses on the study of complex systems that consist of more than one molecule, where order originates from the weak, non-covalent binding interactions between different chemical building blocks [2, 3]. The kinetic and thermodynamic control of covalent bonds has become a challenge for the synthetic community in order to create discrete molecules performing specific functions. This has accelerated chemists to attain precise control over kinetic and thermodynamic courses utilizing weaker inter- and intramolecular interactions, such as hydrogen bonds, van der Waals forces, dipole-dipole interactions, etc. Controlling these weak interactions allows for targeted architectures of new class of molecules containing distinctive kinds of chemical bonds also known as "mechanical bonds." Assemblies derived by the aforementioned forces consist of a distinct number of molecular components that explore mechanical-like movements (output) in response to pre-definite stimulation (input) [4]. The expression is often more generally applied to molecules that modestly mimic functions that occur at the macroscopic level. After organization and assembly, they are capable of linking molecular motions and reactions to complex macroscopic functions including actuation and signal modulation enabling "molecular machines." The combination and coordination of organic, inorganic, and supramolecular chemistry made it possible to build various mechanically interlocked molecular architectures (MIMAs). The field of interlocked molecules is immense, and up to date, research in this field receives high interest and attention. In 2016, Jean-Pierre Sauvage, Sir J. Fraser Stoddart, and Bernard L. Feringa were awarded the Nobel Prize in Chemistry for the "design and synthesis of molecular machines." The term has become state of the art in nanotechnology where a number of favorably complex molecular architectures have been investigated intended to construct MIMAs, estimated to fuel the cutting-edge miniaturization of multifunctional devices (electrical, optical, and chemical) in the near future. The profound investigation of these architectures is endeavored to proceed rapidly due to their valuable properties and potential future applications in biomechanics, molecular electronics, catalysis, drug delivery, electronic materials, and sensing including in general the targeted design of smart novel materials. Photoconductive multifunctional materials involving interlocked molecules and macromolecules are of high importance as they might result in novel hi-tech applications spanning from solar cells and molecular photodiodes to sensing biological applications. In this chapter, we provide a review to published photoconductive interlocked molecules and macromolecules, and we indicate the potentials of various classes of interlocked organic photoconductive dyes.

2. Rotaxanes and catenanes

Molecular machines can be divided into two main categories: synthetic and biological. Large, synthetic molecular machines refer to molecules that are artificially designed and synthesized, whereas biological molecular machines are going deep back in history and can be found under various forms in the nature (transport proteins such as kinesin, myosin, and dynein) [5]. Following a self-assembly process, the formation of large molecular and macromolecular structures can be achieved. These assemblies are mainly interlocked, and no covalent bond is responsible for their stability. Stabilizing interaction may be (i) donor/acceptor forces, (ii) metal/ligand coordination, (iii) hydrogen bonding interactions, (iv) π–π stacking, (v) solvophobic repulsion, and/or (vi) electrostatic forces. Non-covalent interactions enable new properties and smart functional materials by the emerging synergy between molecular recognition and advanced chemistry. The introduction of a mechanical bond enters within the well-recognized chemistry of the subcomponents of supramolecular architectures such as catenanes and rotaxanes. Catenanes and rotaxanes are among the simplest examples of mechanically interlocked molecules with nanometer-scale structures [6]. Many of these molecular assemblies constitute nanomaterials that have been intensively investigated because of their ability to act as molecular machines and/or switches by giving controllable and reversible transformations. The interlocked components can be forced through a combination of chemical, optical, or electrochemical stimuli to change their orientation with respect to one another [7]. These reversible transformations may exhibit high response rates to various highly controllable physical or chemical external stimulations such as pH changes, electricity, light irradiation, heating or cooling, etc. Rotaxanes and catenanes are promising systems for the construction of artificial molecular machines. Catenanes were among the first supramolecular structures that have been reported, in which two or even more cyclic molecules have been mechanically interlocked together and did not disassembly by any external stimuli [4]. In 1964, the first catenane was synthesized by Schill and Lüttringhaus [8]. Synthetic strategies were improved in the late 1980s and beginning of 1990s, in large extent by Stoddart and coworkers [1]. According to the IUPAC nomenclature, [n]catenanes consist of n-interlocked rings. In their simplest form ([2]catenanes), two rings are non-covalently bound forming a structure like the one depicted in **Figure 1** [11]. One of the most synthetically challenging examples of catenanes has been reported in 1994 by the group of Stoddart which was composed of five interlocking macrocycles representing a [5]catenane also known as olympiadane [12].

The disassembly of catenanes into its individual chemical components requires the breaking of one or more covalent bonds within the mechanically linked molecule. One of their fascinating functions is their ability to act as molecular machines where within these assemblies one or more of the macrocyclic ring(s) change position with respect to one another [13]. High synthetic challenges surround the synthesis of catenanes since a macrocyclization reaction is required in order to achieve the interlocked architecture with attendant competition between cyclization and oligomerization. In order to overcome this challenge, catenanes are typically formed under highly diluted conditions which lead consequently to prolonged reaction times,

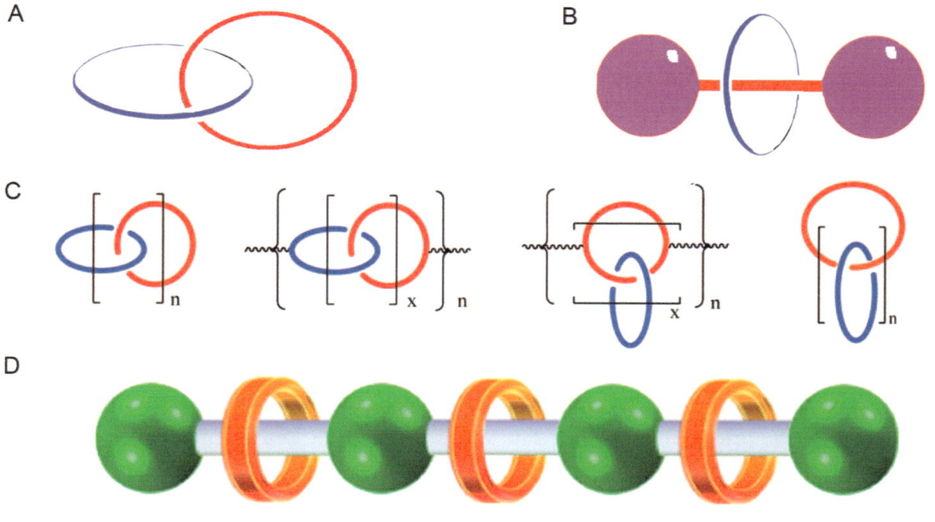

Figure 1. Symbolic representations of a [2]catenane (A) and a [2]rotaxane (B). (C) Various types of polycatenanes. Reprinted with permission from Niu and Gibson [9]. (D) Topological alignment in a polyrotaxane. Reprinted with permission from Yu et al. [10].

since the association between the ring and a macrocycle precursor is weak which diminishes yields. Immense amount of work has been published altering the synthetic protocols as well as introducing a variety of macrocyclic molecules, e.g., crown ethers, cyclophanes, cyclodextrins (CDs), cucurbituril, calixarene, etc. In contrast to catenanes, rotaxanes are composed of one or more macrocycles and "dumbbell-shaped" molecule(s) threaded through them. Stoppering bulky end groups also called "stoppers" prevent disassembly [14]. The word rotaxane is derived from the Latin words "rota" meaning wheel and "axis" meaning axle. The formal naming of rotaxanes according to IUPAC rules is [n]rotaxane, where n indicates the number of chemically independent components in a rotaxane assembly. The simplest form is "[2] rotaxane" which consists of one macrocycle and one dumbbell-shaped molecule. There are several interactions that can initiate self-assembly, needed for the formation of these supramolecular structures in a more efficient manner. These interactions may be hydrophobic, hydrogen bonding, or donor-acceptor interactions. The strength of these interactions varies, and this introduces different stability of the formed complexes depending on the nature of interaction, e.g., Van der Waals forces (2–4 kJ/mol), hydrophobic interactions (4–12 kJ/mol), and hydrogen bonds (8–40 kJ/mol). When considering the strategies of chemical synthesis of rotaxanes, one can distinguish three general approaches: Strategy I, threading of a macrocycle onto a rod molecule and subsequent interaction of the complex formed with the blocking reagents; Strategy II, cyclisation in the presence of compounds having a dumbbell-like structure; and Strategy III, temperature-induced "slipping" of the macrocycle onto bulky terminal groups of the dumbbell-shaped molecule. Accordingly, mechanically interlocked rotaxanes constitute some of the most appropriate candidates to serve as molecular switches and machines in the rapidly developing fields of nanoelectronics and nanoelectromechanical systems (NEMS). Numerous organic cyclic host compounds such as donor-acceptor complexes [15], crown ether complexes [16], and hydrogen bonded complexes involving cyclic amides [17] have been used for rotaxane synthesis. Herein, we shall focus on various rotaxanes and catenanes exhibiting photoconductive properties.

3. Photoconductivity

Photoconductivity is the phenomenon in which electrical current is generated in materials under light radiation. When a material such as a semiconductor absorbs photons of sufficient energy, the electrons in the valence band can be excited, cross the bandgap, and lead to the formation of electron-hole pairs resulting in increased conductivity (**Figure 2**). In principle, photoconductivity is a common physical phenomenon of a material, and it is particularly prominent in semiconductors due to their small bandgaps. Thus, photoconductivity generates great interest for the investigation of the electronic structure, transportation properties of materials, electron-hole pair dynamics, as well as practical applications such as photodetectors, photoresistors, and charge-coupled devices. The classic photoconductive materials consist of doped semiconductors, e.g., Si, Ge and Se [18–20], metal oxides, and sulfides as well as conductive polymers. Apart from these classic materials, the photoconductivity is also observed in an ultracold fermionic gas that is trapped in an optical lattice [21] as well as various organic compounds [22].

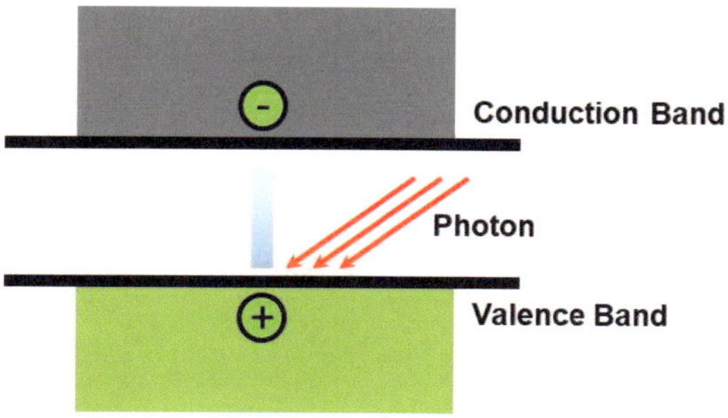

Figure 2. Excitation process leading to photoconductivity in a condensed matter system, in which incoming laser light ($h\nu$) excites an electron to the valence band, leaving a positively charged hole behind.

4. Photoconductivity measurements

The most prominent method to measure the photoelectrical properties is xerography (method shown schematically in **Figure 3**) [22]. The target sample is mounted and grounded on a sample holder, which can move forward or backward through a driving chain (**Figure 3a**). When the sample is moved to position (2) where the corotron is just above, the sample can be charged either positively or negatively. When the sample is moved to position (3), its surface potential can be measured by using an electrometer. A typical scheme of a photodischarge curve produced using this method is shown in **Figure 3b**. When the electronic shutter is closed, the sample is under totally dark conditions, and dark conductivity can be measured. When the shutter is open, the sample can be under exposure of either an intense erase light to measure the residual potential or a monochromatic light with known intensity to measure the photosensitivity.

This technique is simple and allows for the determination of first-order xerographic properties. Therefore, it has been widely used in the study and evaluation of photoconductive organic material properties [22]. Another intensively utilized method to investigate the photoconductivity of the material is graphically shown in **Figure 4**. By illuminating the sample with light of various wavelengths and plotting the current evolutions as a function of time, the generated photocurrent can be measured as well as different photoeffects [23, 24], e.g., photodoping,

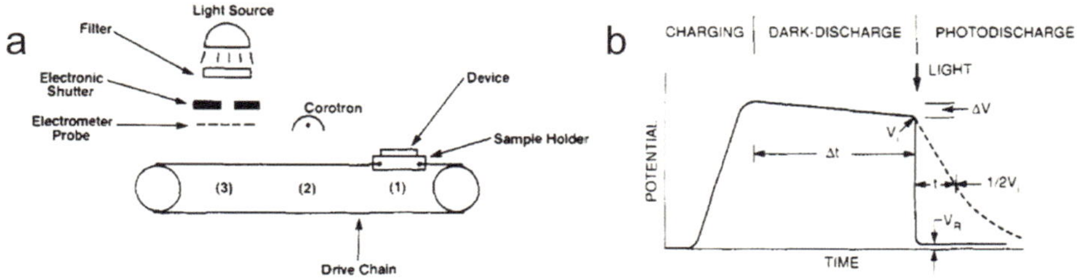

Figure 3. A schematic of the apparatus (a) and photodischarge curves (b). Reprinted with permission from Law [22].

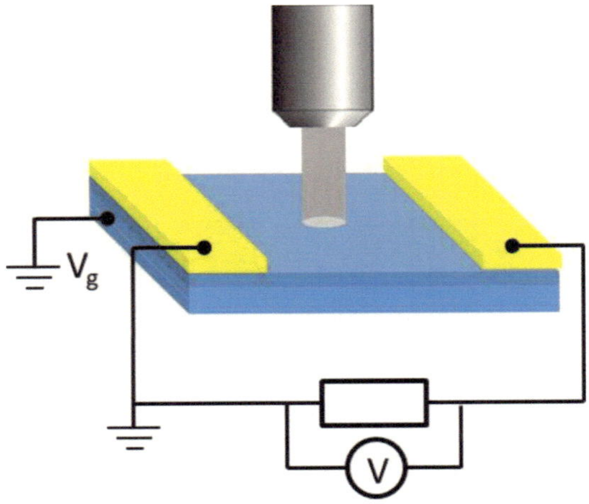

Figure 4. A schematic of photoconductance experimental setup.

photogating, etc. can be well studied. In addition, by tuning the back-gate voltage of the device, more phenomena such as photoconductive gain effect can also be studied [25].

5. Photoconductive interlocked molecules

5.1. Photoconductive rotaxanes and polyrotaxanes

5.1.1. Why rotaxanes?

Rotaxanation, i.e., the inclusion of an axial molecule in the cavity of a macrocyclic molecule, is an interesting approach for the design of novel photoconductive materials which can efficiently introduce a number of beneficiary characteristics to these materials. There are different reasons which could justify why designing rotaxane photoconductive structures can lead to promising new materials. First of all, the moieties or functional groups which introduce photoconductivity to a compound are often unstable and chemically labile. The cavities of suitable macrocycles could offer protection to such entities, and this is vital for the durability and proper function of a photoconductive compound or material. That is, for instance, the case of azo dyes and squaraines, the rotaxanes of which will be examined in this chapter. Moreover, in many cases of photoconductive materials, prevention of intermolecular interactions is sought after. Encapsulation of photoconductive axial molecules in molecular rings often reduces the intermolecular interactions such as π–π stacking without hampering the charge transport. This is especially important in photoluminescent compounds where parallel alignment and interaction of π-conjugated molecules are obstacles. In addition to that, the non-covalent interactions developed between the axial and macrocyclic components in a rotaxane are overly important as they provide geometry stabilization and optimal orientation of these components so that charge transfer and transport are efficient. Such interactions are essential for the photoconductive behavior of a number of viologen-involving rotaxanes bearing electron-donating counterparts; a few such systems are reviewed herein. Furthermore, the encapsulation of long conductive/photoconductive macromolecules within macrocycles such as CDs could result in the development of polyrotaxane molecular wires which, in a similar fashion to their macroscopic wire analogues, possess a conducting internal and insulating external part. The insulating part could prevent short-circuit problems in future molecular circuits involving these wires. The role of the insulating macrocycles in the photoconductivity of polyrotaxanes is also reviewed in this chapter. The last but substantially beneficiary feature of the development of photoconductive rotaxanes and polyrotaxanes is multifunctionality. Up to date a vast number of rotaxanes have been reported undergoing fully controlled shuttle motions, exhibiting switchability, photo- and electro-chromic, and photoluminescent properties. Combining one or more of these promising properties with photoconductivity could result in novel types of materials able to perform multiple functions upon demand.

5.1.2. Azo dyes and rotaxanes thereof

Azo dyes constitute a widely known class of organic pigments with significant industrial interest. These dyes exist in numerous products of everyday use, ranging from textile and leather dyeing agents to food colorants and DVD−R/+R disc recording layer materials. All azo compounds contain one or more units of the azo (–N=N–) chromophore connected to carbon atoms in both sides. The vast majority of azo dyes bear an azo group coupled to aromatic substituents such as benzene or naphthalene rings. To date a remarkable number of azo dyes have been synthesized and characterized [26]. This large number comes as a consequence of the ease of synthesis of azo compounds mainly relying on azo coupling, which involves an electrophilic substitution reaction between an aryl diazonium cation and a coupling partner. Typical coupling partners can be various aromatic compounds possessing electron-donating groups such as –OH, –OR, or $–NR_2$ functional groups [27]. Numerous rotaxanes comprising the arylazo units have been also reported [26].

The properties of azo dyes are not merely related to their color/light absorption properties but also to their vivid photochemistry, as they readily undergo reversible E/Z isomerizations (see **Figure 5**) in most of the cases via excitation with near UV or even visible light [28]. The reversibility as well as the low light energy demand for the accomplishment of this photochemical process renders azo compounds and materials thereof even more attractive for a number of optoelectronic applications [26–28]. Indeed, this photoreaction has been exploited in plenty azo-bearing materials including rotaxanes. Murakami et al. described 20 years ago the first light-driven molecular shuttle based on an azo rotaxane [29]. Deligkiozi et al. have reported controllable shuttling motions of α-cyclodextrin in [2]rotaxanes bearing a fully conjugated arylazo-based linear part [30]. Tian and coworkers reported on an azo-involving light-driven rotaxane molecular shuttle with dual fluorescence addresses comprising two different fluorescent naphthalimides and α-CD [31]. Indeed to date numerous other examples of photosensitive azo-involving rotaxanes have been reported [32]. Importantly, many azo compounds are known to exhibit photoconductive behavior. In 1969, Rau was the first to report the photoconductive behavior of azo dyes and specifically observed the photocurrents that form thin layers of a simple azo compound: l-(phenylazo)-2-naphthol (**Figure 5**) [33]. Six years after the pioneering work of Rau, Champ and Shattuck reported the use of chlorodiane blue, a bis-azo compound (a derivative of 1-(phenylazo)-2-naphthol) as a photogenerating pigment in xerographic devices [34]. These two early scientific reports initiated a huge endeavor for the

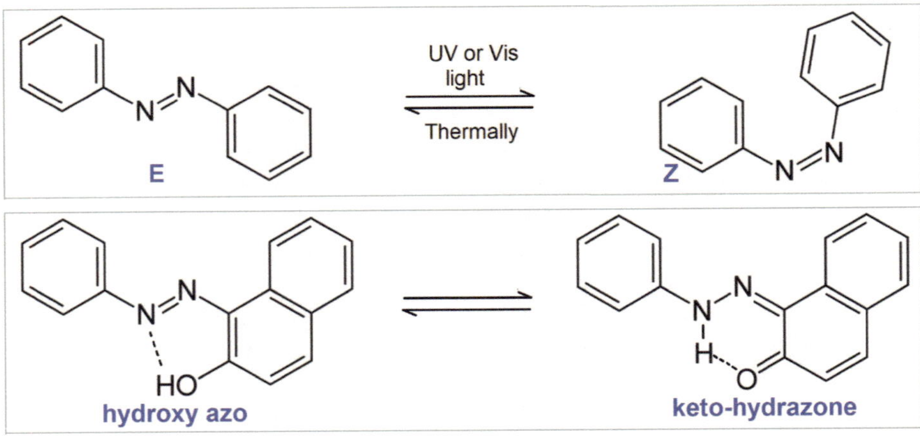

Figure 5. Scheme depicting the reversible E/Z isomerization of azobenzene (upper panel). The hydroxyl azo/ketohydrazone tautomerism of l-(phenylazo)-2-naphthol (lower panel).

development of novel azo pigments with photoconductive properties, an endeavor which continues to date. Many research groups have come out with various photoconductive azo compounds mostly with structures relative to the parent l-(phenylazo)-2-naphthol, over the years [22]. The photoconductive behavior of this parent azo pigment is narrowly connected to its structure and specifically to the hydroxyl azo/ketohydrazone tautomerism that this molecule and its derivatives exhibit (**Figure 5**) [22].

Nonetheless, there are also recent reports of photoconductive azo dyes with structure different from the "inspiring" structure of l-(phenylazo)-2-naphthol. Recently, Deligkiozi et al. observed photocurrents from a [2]rotaxane of an azobenzene-based dye encapsulated in α-cyclodextrin (α-CD) as shown in **Figure 6** [35]. The photoconductivity of this interlocked azo dye measured using a wet method [35] was proved to be significantly higher than its α-CD-free precursor. The aforementioned [2]rotaxane was one of the first examples of rotaxanes involving an axial linear part with full π-conjugation [30, 36]. This robust aromatic skeleton provides the α-CD-free precursor some conductivity even in the dark which is reinforced when the dye is irradiated with white light. Remarkably though, the corresponding [2]rotaxane with α-CD appears to exhibit a significantly higher photoconductivity than the α-CD-free dye. Here, it is important to note that supramolecular insulation provided by α-CD (an insulating compound) is expected to result in a reduced conductivity of the [2]rotaxane when compared to its α-CD-free analogue. Yet, Cacialli et al. have shown that CD-encapsulated conductive polyrotaxanes with poly(para-phenylene) and poly(4,4′-diphenylene vinylene) continue to exhibit high conductivity despite the cyclodextrin insulating impact. It was concluded that cyclodextrin encapsulation inhibits parallel arrangement of the molecules without causing elimination of charge transport [37].

In the case of [2]rotaxane by Deligkiozi et al., photoconductivity was rationalized in terms of the non-covalent interactions of the cavity of α-CD and the encapsulated part of the azobenzene unit of this compound. These interactions result in stabilization of the geometry of the azobenzene part of the molecule [35]. Presumably, this stabilizing geometry effect resembles the corresponding effect observed in the case of the tautomeric l-(phenylazo)-2-naphthol derivatives (**Figure 5**).

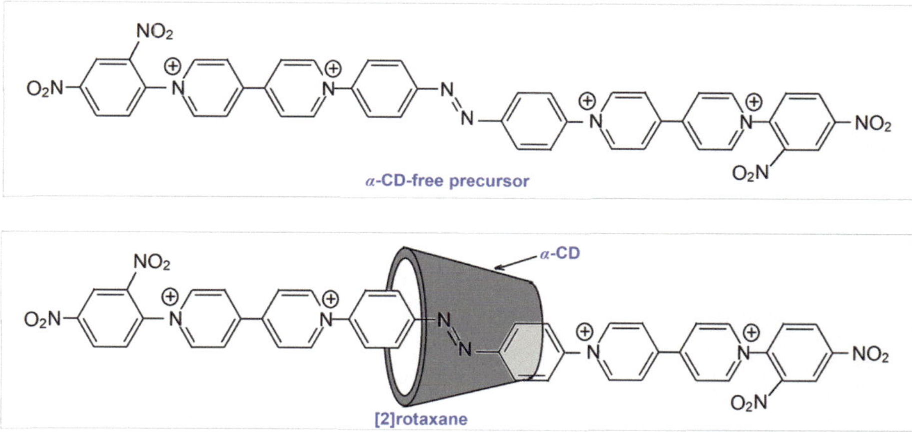

Figure 6. The chemical structure of the tetracationic part of the [2]rotaxane by Deligkiozi et al. [35] (lower panel) and that of the tetracationic linear α-CD-free precursor (upper panel).

The supramolecular insulation provided by encapsulation of an azo dye in α-CD has been earlier utilized by Haque et al. [38]. In their work they managed to thread π-conjugated tri-azo dye molecules through α-CD and then immobilize the resulting [2]rotaxanes onto nanocrystalline TiO_2 films (**Figure 7**). Transient absorption spectroscopy experiments supported that charge recombination was considerably retarded in the case of the as formed TiO_2 films when compared to non encapsulated dyes. This finding is very stimulating as it indicates that photocurrents are still generated by the conjugated encapsulated molecules, while the insulating α-CD part maintains a slow charge recombination. In the light of that, these photoconductive interlocked azo compounds are considered as promising for dye-sensitized solar cell (DSSC) applications [39].

5.1.3. Viologen-involving rotaxanes

Viologens constitute a class of heterocyclic compounds with remarkable properties [40]. They are 4,4′-bipyridine derivatives having both their nitrogen atoms quaternized, i.e., they are substituted by a chemical group which is often an aliphatic chain or an aryl group (see **Figure 8**) [40]. Due to their intense electron withdrawing (EW) character, aromaticity, as well as photo- and electro-active nature, they have been utilized as key components in a vast number of new materials [40, 42]. Viologens are also well known for their intense electrochromism which is attributed to the reversible one-electron reduction they readily undergo electrochemically or by means of reducing agents. They readily form charge transfer complexes (CTCs) with a variety of electron-donating species, e.g., ferrocyanides [43], tetrathiafulvalene (TTF) derivatives [41], as well as phenols [44]. In these complexes charge is reversibly transferred from the electron-donating part to the viologen upon absorption of visible light. Because of that, CTCs are colorful compounds and very photosensitive. Today, there is clear evidence that CTCs involving viologens exhibit photoconductive properties. This is, for instance, the case in some recent reported viologen/TTF CTCs (see **Figure 8**). Huo et al. observed marked photocurrent responses directly from such crystalline CTCs or from

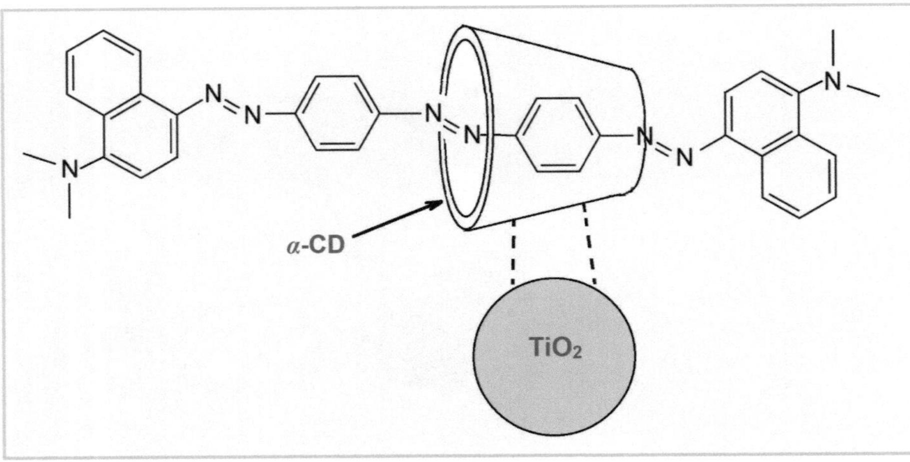

Figure 7. Schematic representation of the α-CD-encapsulated tri-azo dye onto TiO_2 nanocrystalline films utilized by Haque et al. (see Ref. [38]).

prepared film electrodes involving the CTCs depicted in **Figure 8**. A large number of rotaxanes involving viologens have also been reported exhibiting donor-acceptor interactions in which viologens play an important role as strong EW species [32, 45]. In these rotaxanes, viologen units are encountered either as parts of the axial or as parts of the macrocyclic components. In the latter class of rotaxanes, they are often used in cyclophane structures (see **Figure 9**). Such rotaxanes are considered for high-tech applications due to the ease of control of their function through electrical or light triggering. Feng et al. have achieved reproducible nanorecording on rotaxane thin films comprising TTF-involving axial and viologen cyclophane components [47]. A few years ago, Sheeney-Haj-Ichia and Willner reported that pseudorotaxane monolayers comprising viologen cyclophane units exhibit photocurrents eightfold higher than the ones observed in the case of the control monolayers lacking the viologen component (**Figure 9**) [46].

These significantly amplified photocurrents observed in the pseudorotaxane assembly were attributed to vectorial electron transfer of photoexcited conduction-band electrons to the strong electron accepting component. According to the authors [46], this fact leads to charge separation and retardation of electron-hole recombination. This finding is also in line with the photoconductive character of viologen CTCs and indicates that interlocked molecules and macromolecules comprising viologen CTC entities are promising materials with potential photoconductive properties.

In 2007 Saha et al. [48] reported on a redox-driven multicomponent rotaxane shuttle comprising a linear component which involved TTF, a naphthyl ether, and a porphyrin. The macrocyclic component employed was the same bis-viologen cyclophane utilized by Sheeney-Haj-Ichia and Willner [46]. C_{60} was utilized as a strong electron acceptor/bulk ending group. The authors emphasized that such donor-chromophore-acceptor system could generate photocurrents. This example constitutes one interesting case of a multifunctional material with potential photoconductive properties being able to also function as a molecular shuttle.

http://dx.doi.org/10.5772/intechopen.79798

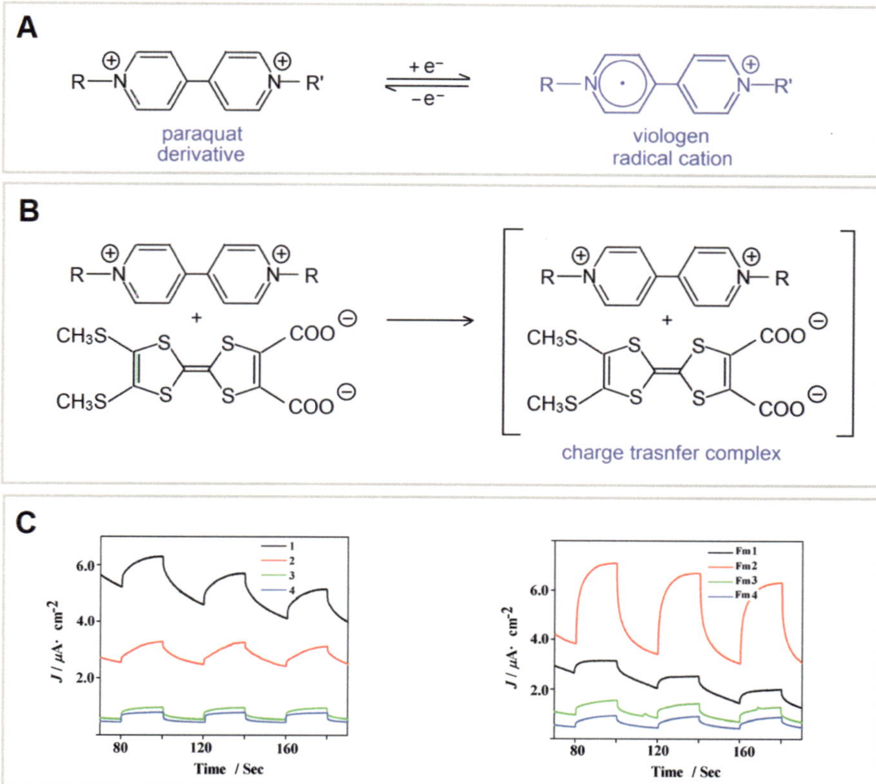

Figure 8. (A) Scheme representing the reversible one-electron reduction of a viologen dication comprising two different substituents (R and R′). (B) Scheme depicting the formation of photoconductive CTC complexes of a group of viologens and a dianionic TTF derivative according to Huo et al. [41]. (C) The photocurrent responses observed from the CTCs of middle panel, measured in crystals of the CTCs (left) and in thin-film electrodes (right). Plots of panel (C) reprinted with permission from Huo et al. [41].

5.1.4. Squaraine rotaxanes

Squaraine compounds constitute a widely known class of organic photoconductive compounds [22]. It was as early as 1966 when Sprenger and Ziegenbein reported the synthesis of intensely colored compounds derived from squaric acid (see Structure **I** in **Figure 10**) [49]. It was observed that the compound produced is characterized by a unique electronic structure resulting in interesting properties. Many relative compounds were subsequently synthesized. These fascinating compounds bear an internal donor-acceptor-donor (D-A-D) structure which can be represented through the resonance structures depicted in **Figure 10** [22, 50]. Around 40 years ago, Schmidt proposed the name squaraine for these compounds [22]. In 1974 Champ and Shattuck were the first to report the photoconductive properties of squaraine dyes [51]. They revealed that squaraines are able to generate electron-hole pairs in bilayer xerographic devices through light irradiation [51]. Awhile before this report, squaraines had already been proposed as sensitizers for ZnO photoconductors [52]. As mentioned squaraines are deeply colored compounds, and their absorption and emission are situated in the deep-red and near-infrared (NIR)

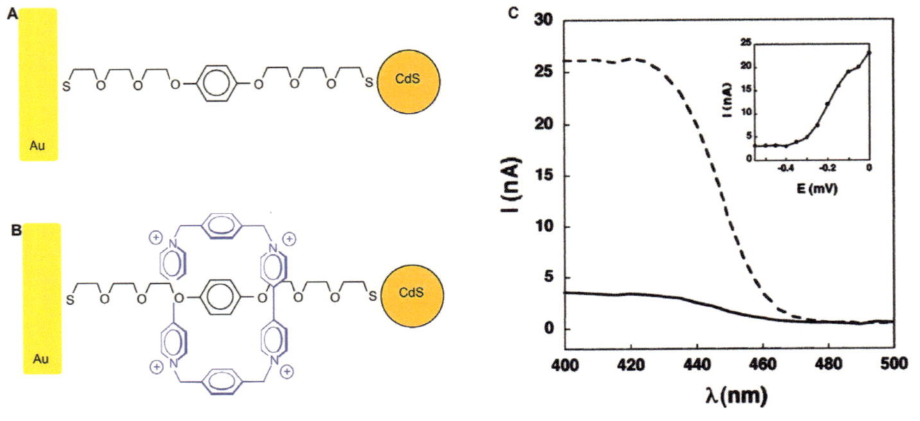

Figure 9. (A) Illustration of the setup used by Sheeney-Haj-Ichia and Willner without cyclophane. (B) The setup after inclusion in a tetracationic cyclophane. (C) Plot depicting the photocurrent response observed for the system in panel A (solid line) and that in panel B (dashed line) vs. the irradiation wavelength. Plot of panel (C) reprinted with permission from Sheeney-Haj-Ichia and Willner [46].

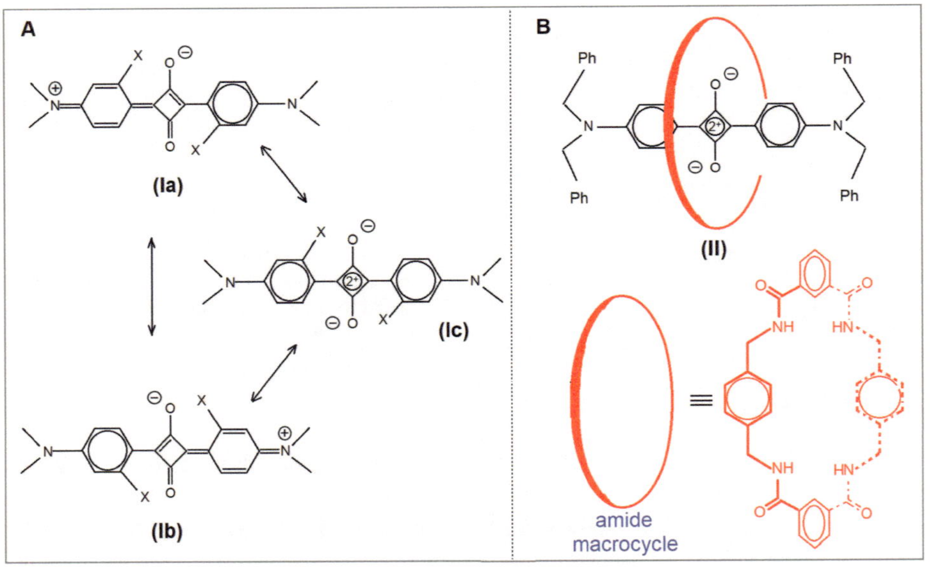

Figure 10. (A) The resonance structures of squaraines (**Ia–c**). (B) A squaraine rotaxane involving an amide macrocycle.

region [50]. These features along with their photoconductivity render squaraines important candidates for DSSC applications. In these technologies novel sensitizers absorbing in NIR wavelength region are required in order to boost the photoconversion efficiency. Indeed Yum et al. reported a photoconversion efficiency as high as 4.5% when using an unsymmetric squaraine dye (structure **III** depicted in **Figure 11B**) [54]. This work essentially indicated that squaraines are useful candidates for DSSC (details in **Figure 11**). Apart from marked photoconductive compounds, squaraines are generally very photosensitive and fluorescent [50].

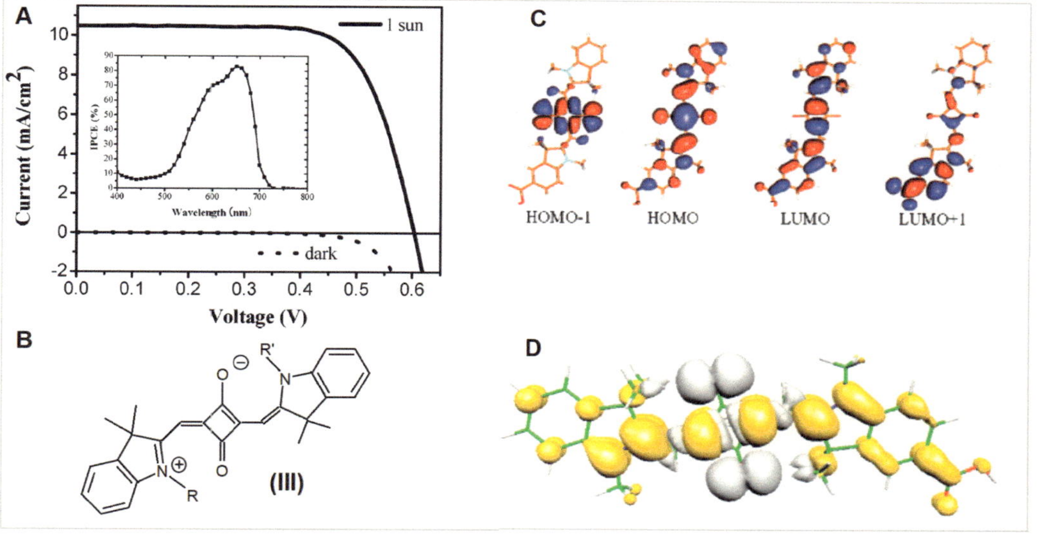

Figure 11. (A) Current-voltage characteristics for a squaraine dye (type (II) with R = C_8H_{17} and R′ = Et) and photocurrent action spectrum (inset). (B) Chemical structure of squaraine **III,** (C) frontier orbitals, and (D) charge density of dye **III**. Figures in Panels A, C, and D reprinted with permission from [53].

All these features are narrowly connected to their electronic structure, and they are essential for a vast number of imaging applications [50]. A significant drawback of squaraines is their instability against strong nucleophiles as well as their aggregation propensity which pulls down their fluorescence and potentially photoconductivity. These problems can be solved by the use of protecting threading macrocycles, i.e., through rotaxanation of the sensitive core. This approach was first employed by Leigh and coworkers who managed to synthesize [2] rotaxanes utilizing normal squaraine structures and suitable amide-macrocyclic compounds (**Figure 10B**) [55, 56]. The as structured rotaxanes are characterized by significantly higher chemical and photophysical stabilities than the non encapsulated squaraines. This revolutionary study inspired a lot of other research groups to design and synthesize a wide variety of squaraine-based rotaxanes with potentials in a number of applications [50]. The corresponding rotaxanes do not hamper the properties of squaraines, but instead the properties are retained or even improved. Due to their high photoconductivity, promising performance in DSSC applications as well as other biologically relevant applications of rotaxanes of squaraines, their use is currently seriously considered.

5.1.5. Photoconductive polyrotaxanes

In recent years there is an increasing interest in the design and synthesis/fabrication of molecular wires, i.e., conductive conjugated polymers of high conductivity. Even though the research endeavors to develop molecular wires were initiated theoretically already in the 1940s using quantum mechanics [57], there is today a tremendous interest in this type of nanosized wires for a range of high-tech applications. In such systems prevention of short circuits could be achieved through threading of a conductive polymer within the cavities of insulating (protecting) macrocycles [58]. These polyrotaxane-structured wires also called insulated molecular wires (IMWs), with nanometer dimensions, could be used in nanosized circuits [53]. The role of the insulating components (usually α- and β-CDs) is an important research subject as it clearly affects the conductivity and photoconductivity of polyrotaxane wires. In 2009 Terao et al. [59] studied a permethylated α-CD (PM-α-CD) polyrotaxane of a poly(phenylene ethynylene)-based polymer (**Figure 12A**) and reported the formation of a prominently insulating organic semiconductor wire exhibiting remarkably high hole mobility along the core π-conjugated polymer. They also reported light-induced currents observed upon excitation at λ = 355 nm (**Figure 12A**). Terao et al. some years later [61] based on previous theoretical publications compared experimentally the charge mobilities of linear and zig-zag polyrotaxanes involving conjugated polymers and permethylated α-CD.

They reported increased charge mobilities for the zig-zag polymer and confirmed the light-induced formation of charge carriers in the case of the linear polyrotaxane. However, they observed that rapid free carrier-formation processes were overlapped in the zig-zag polyrotaxanes. These stimulating findings indicate that IMWs do exhibit photoconductivity, but clearly the geometry of the macromolecules affects their photoconductive behavior. Encapsulation of a conjugated polymer such as the aforementioned π-conjugated polymer in the insulating cavities of PM-α-CD leads to increased lifetimes of charged radicals on the conjugated core via hindering charge recombination processes [62]. Moreover encapsulation results in marked fluorescence enhancement in this kind of polyrotaxanes, particularly in the solid state, suggesting that encapsulation is crucial for the achievement of efficient fluorescence properties [62]. More recently, Kostromin et al. [60] studied the photovoltaic effect and charge carrier mobility of some bithiophene conducting polymers, both "bear" and encapsulated in β-CD units (see **Figure 12B**). They concluded that the β-CD introduced marked insulation of

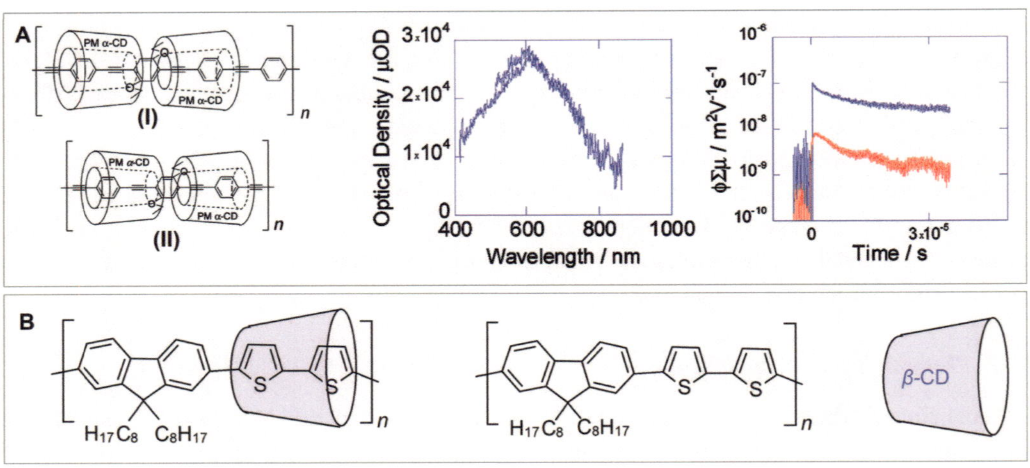

Figure 12. (A) Structure of the IMWs studied by Tarao et al. [59] along with the transient absorption spectrum of IMW (I) after pulse exposure and conductivity transients observed for (I) (blue) and (II) (red) upon 355 nm excitation. Figures reprinted with permission from [59]. (B) Structure of the conjugated polymer and IMW investigated by Kostromin et al. [60].

thiophene fragments of the macromolecule, and this lead to hampering the transport of carriers which in turn yielded in a limited photovoltaic effect [60].

5.2. Catenanes and polycatenanes

5.2.1. The structure of catenanes and its benefits

Catenanes constitute another important class of interlocked molecules. Just like rotaxanes they are stabilized through mechanical bonds [9, 32]. Yet, they consist of two or more macrocycles interlocked in a way that resembles the connectivity of rings in a chain (Latin: *catena* = chain). There is a large variety of catenated structures reported to date with numerous applications. There are various reasons why catenanes could become important candidates for new photoconductive materials. As also mentioned for rotaxanes, encapsulation of a sensitive photoconductive moiety or functionality in a molecule could significantly increase the durability of the material and protect the desired photoconductive properties. Thus, interlocking photoconductive (or more generally photosensitive) macrocycles could potentially lead to promising stable catenated materials with optoelectronic applications. Moreover, geometry fixation and proper orientation in catenanes can give rise to intermolecular interactions (e.g., π–π stacking, etc.) facilitating efficient charge transfer in such materials. This is a key property which is discussed in more detail in this chapter. Finally, polycatenanes involving photoconductive parts could be perfect candidate multifunctional materials, as in such structures one can introduce photoconductivity via embedding repeated photoconductive catenane units in macromolecules with special properties, e.g., electrical or thermal conductivity, mechanical strength, etc.

5.2.2. Photoconductive catenanes

Even though numerous examples of catenanes and polycatenanes have been reported, there is a limited number photoconductive catenanes and polycatenanes. However, there is strong indication that such materials could also exhibit promising photoconductive behavior. The main types of organic photoconductive molecules utilized in rotaxanes and polyrotaxanes can be also utilized in catenanes and polycatenanes.

About 15 years ago, Simone [63] reported on the synthesis and characterization of some polycatenane repeated units of cyclophane connected to thiophene rings (red-colored part in **Figure 13**) interlocked with a bis-viologen tetracationic cyclophane (blue-colored species in **Figure 13**). This approach involving the aforementioned two cyclophanes was initially employed by Stoddart and coworkers [15] and is a very popular combination for numerous rotaxanes and catenanes (see, for instance, the pseudorotaxane in **Figure 9**). The resulting polycatenane of Simone and Swager is stabilized through π-stacking between the aromatic bipyridinium and benzene-1,4-diether rings of the interlocked macrocycles. The catenane in **Figure 13** which is colored green exhibits a charge transfer visible band situated at λ = 626 nm. This polycatenane as well as another variant was reported to be conductive (linear part is a π-conjugated polymer) [64] but also to exhibit significant photocurrent responses [63]. This example constitutes an important case enabling the design of novel photoconductive polycatenanes.

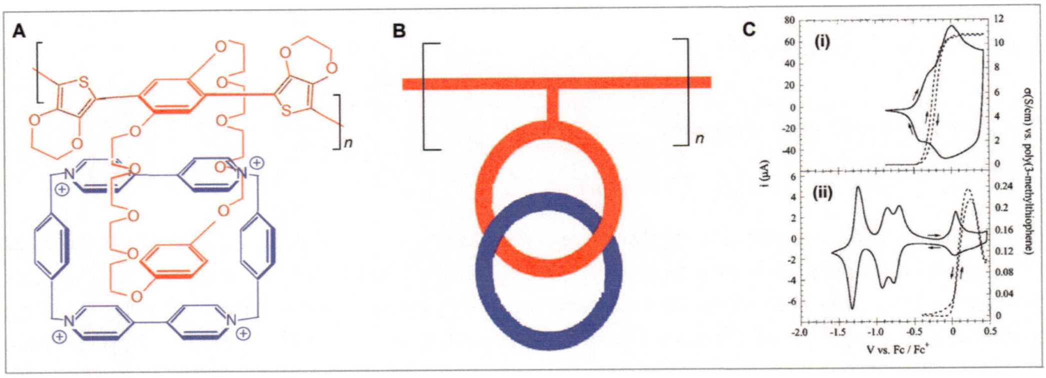

Figure 13. (A) The polycatenane synthesized by Simone [63]. (B) Symbolic representation of the polycatenane in panel A. (C) Cyclic voltammograms (solid lines) and conductivity profiles (dashed lines) for the polymer lacking the cyclophane units (colored blue) (i) and polycatenane (ii). Panel C plots reprinted with permission from [64].

It is very important to stress that (poly)catenanes do not exhibit disadvantages when compared to (poly)rotaxanes in terms of their photoconductive behavior/properties. The "strategies" for photocurrent generation are essentially the same for both classes of interlocked (macro)molecules. The downside in the case of (poly)catenanes can sometimes be the more tedious synthetic methodology required, when compared to (poly)rotaxanes (see paragraph 2). To some extent, this might explain the limited number of reported photoconductive (poly) catenanes. Nevertheless, catenated structures are certainly capable of introducing stability and shielding of the photoconductive parts. Additionally, catenated structures could potentially maintain efficient photocurrent generation and slow charge recombination in photoconductive materials. Thus, they should be considered as promising photoconductive interlocked materials/compounds, and they should clearly be given more attention.

6. Applications of rotaxanes and catenanes

Rotaxanes and catenanes are gaining more and more attention due to their applicability in modern technologies. They have been proposed for numerous biological applications such as smart drug-delivery systems corresponding to anticancer drugs [65, 66], imaging of biological matter (e.g., mitochondria) [67], or as useful materials for the enhancement of MRI imaging [68]. Especially, the squarain-involving interlocked molecules described are prominent examples of fluorescent bio-imaging agents and chemosensors [50]. Furthermore, both types of interlocked molecules are prominent candidates for new smart future applications acting as (multi)functional materials and undergoing fully controllable switching, shuttle motions, as well as molecular motor functions [69–71]. Medium- and photo-responsive interlocked molecules are also currently considered as new sensing materials with various possible applications [30, 35, 36, 72]. Taking into account the potentials of the photoconductive interlocked molecules described in this chapter, one could foresee a bright future for new optoelectronic materials, molecular wires, photoconductors, photovoltaics, and many other novel applications. Especially, (poly)catenanes and (poly)rotaxanes encompassing CTC units are of high importance as materials with significant photoconductivity and photosensitivity. It is high time this fascinating class of interlocked (macro)molecules was given more attention.

7. Conclusion

This chapter has provided a review of the research field of interlocked molecules and macromolecules placing emphasis on rotaxanes, catenanes, and polymeric structures thereof. Various categories of organic photoconductive rotaxanes and catenanes have been reviewed, and the main structural and photoconductive characteristics have been provided. The (photo) conductive properties of the molecules and macromolecules with and without encapsulation are compared. A range of examples and potential applications has been also provided.

Author details

Raffaello Papadakis[1,2]*, Ioanna Deligkiozi[2] and Hu Li[3]

*Address all correspondence to: rafpapadakis@gmail.com

1 Department of Chemistry—Ångström Laboratory, Uppsala, Sweden

2 School of Chemical Engineering, National Technical University of Athens (NTUA), Laboratory of Organic Chemistry, Athens, Greece

3 School of Electrical and Electronic Engineering, University of Manchester, Manchester, United Kingdom

References

[1] Barnes JC, Mirkin CA. Profile of Jean-Pierre Sauvage, Sir J. Fraser Stoddart, and Bernard L. Feringa, 2016 Nobel Laureates in Chemistry. PNAS, 2017;**114**(4):620-625. DOI: 10.1073/pnas.1619330114

[2] Lehn J, Mint M. Perspectives in supramolecular chemistry-from molecular recognition towards molecular information processing and self-organization. Angewandte Chemie. 1990;**29**(11):1304-1319. DOI: 10.1002/anie.199013041

[3] Würthner F. Supramolecular Dye Chemistry. Heidelberg: Springer; 2005. DOI: 10.1007/b105136

[4] Zheng YB, Yang YW, Jensen L, Fang L, Juluri BK, Flood AH, Weiss PS, Stoddart JF, Huang TJ. Active molecular plasmonics: Controlling plasmon resonances with molecular switches. Nano Letters. 2009;**9**:819. DOI: 10.1021/nl803539g

[5] Kay ER, Leigh DA. Rise of the molecular machines. Angewandte Chemie, International Edition. 2015;**54**:10080-10088. DOI: 10.1002/anie.201503375

[6] Browne WR, Feringa BL. Making molecular machines work. Nature Nanotechnology. 2006;**1**:25-35. DOI: 10.1038/nnano.2006.45

[7] Braunschweig AB, Dichtel WR, Miljanić OŠ, Olson MA, Spruell JM, Khan SI, Heath JR, Stoddart JF. Modular synthesis and dynamics of a variety of donor-acceptor interlocked compounds prepared by a click chemistry approach. Chemistry, an Asian Journal. 2007;**2**: 634-647. DOI: 10.1002/asia.200700035

[8] Schill G, Lüttringhaus A. The preparation of catena compounds by directed synthesis. Angewandte Chemie, International Edition. 1964;**3**:546-547. DOI: 10.1002/anie.196405461

[9] Niu X, Gibson HG. Polycatenanes. Chemical Reviews. 2009;**109**:6024-6046. DOI: 10.1021/cr900002h

[10] Yu L, Li M, Zhou XP, Li D. Hybrid inorganic-organic polyrotaxane, pseudorotaxane, and sandwich. Inorganic Chemistry. 2013;**52**:10232-10234. DOI: 10.1021/ic401722c

[11] Safarowsky EO, Windisch B, Mohry A, Vögtl F. Nomenclature for catenanes, rotaxanes, molecular knots, and assemblies derived from these structural elements. Journal für Praktische Chemie. 2000;**342**:437-444. DOI: 10.1002/1521-3897(200006)342:5<437::AID-PRAC437>3.0.CO;2-7

[12] Amabilino DB, Ashton PR, Reder AS, Spencer N, Stoddart JF. Olympiadane. Angewandte Chemie, International Edition. 1994;**33**(12):1286-1290. DOI: 10.1002/anie.199412861

[13] Wilson MR, Solà J, Carlone A, Goldup SM, Lebrasseur N, Leigh DA. An autonomous chemically fuelled small-molecule motor. Nature. 2016;**534**:235. DOI: 10.1038/nature18013

[14] Ashton PR, Baxter I, Fyfe MCT, Raymo FM, Spencer N, Stoddart JF, White AP, Williams DJ. Rotaxane or pseudorotaxane? That is the question! Journal of the American Chemical Society. 1998;**120**:2297-2307. DOI: 10.1021/ja9731276

[15] Sauvage JP, Gaspard G, editors. From Non-Covalent Assemblies to Molecular Machines. 1st ed. Weinheim: Wiley VCH Verlag GmbH; 2011

[16] Fielden SDP, Leigh DA, McTernan CT, Pérez-Saavedra B, Vitorica-Yrezabal IJ. Spontaneous assembly of rotaxanes from a primary amine, crown ether and electrophile. Journal of the American Chemical Society. 2018;**140**(19):6049-6052. DOI: 10.1021/jacs.8b03394

[17] Schalley CA, Weilandt T, Bruggemann J, Vogtle F. Templates in Chemistry I. Vol. 248. Berlin: Springer; 2004

[18] Newman R, Tyler WW. Photoconductivity in germanium. Solid State Phys. 1959;**8**:49-107. DOI: 10.1016/S0081-1947(08)60479-8

[19] Vavilov VS, Lotkova EN, Plotnikov AF. Photoconductivity and infra-red absorption in silicon irradiated by neutrons. Journal of Physics and Chemistry of Solids. 1961;**22**:31-38. DOI: 10.1016/0022-3697(61)90239-6

[20] Park YS, Reynolds DC. Exciton structure in photoconductivity of CdS, CdSe, and CdS: Se single crystals. Physical Review. 1963;**132**:2450-2457. DOI: 10.1103/PhysRev.132.2450

[21] Heinze J, Krauser JS, Fläschner N, Hundt B, Götze S, Itin AP, Mathey L, Sengstock K, Becker C. Intrinsic photoconductivity of ultracold fermions in optical lattices. Physical Review Letters. 2013;**110**:085302. DOI: 10.1103/PhysRevLett.110.085302

[22] Law KY. Organic photoconductive materials: Recent trends and developments. Chemical Reviews. 1993;**93**:449-486. DOI: 10.1021/cr00017a020

[23] Rossler C, Hof KD, Manus V, Ludwig S, Kotthaus JP, Simon J, Holleitner AW, Schuh D, Wegscheider W. Optically induced transport properties of freely suspended semiconductor submicron channels. Applied Physics Letters. 2009;**93**:071107. DOI: 10.1063/1.2970035

[24] Martinez I, Ribeiro M, Andres P, Hueso LE, Casanova F, Aliev FG. Photodoping-driven crossover in the low-frequency noise of MoS2 transistors. Physical Review Applied. 2017; **7**:1-8. DOI :10.1103/PhysRevApplied.7.034034

[25] Hof K D, Rossler C, Manus S, Kotthaus J P, Holleitner A W, Schuh D, Wegscheider W. Dynamic photoconductive gain effect in shallow-etched AlGaAs/GaAs quantum wires. Physical Review B-Condensed Matter and Materials Physics. 2008;**78**:2-6. DOI: 10.1103/PhysRevB.78.115325

[26] Zollinger H. Color chemistry: Syntheses, Properties, and Applications of Organic Dyes and Pigments. 3rd ed. Zürich: Wiley-VCH, Verlag Helvetica Chimica Acta; 2003. ISBN: 3-906390-23-3

[27] Zollinger H. Diazo Chemistry: Aromatic and Heteroaromatic Compounds. Vol. 1. 1st ed. Weinheim: Wiley-VCH; 1994. ISBN: 3527292136

[28] Mahimwalla Z, Yager KJ, Mamiya J, Shishido A, Priimagi A, Barrett CJ. Azobenzene photomechanics: Prospects and potential applications. Polymer Bulletin. 2012;**69**:967-1006. DOI: 10.1007/s00289-012-0792-0

[29] Murakami H, Kawabuchi A, Kotoo K, Kunitake M, Nakashima N. A light-driven molecular shuttle based on a rotaxane. Journal of the American Chemical Society. 1997;**119**:7605-7606. DOI: 10.1021/ja971438a

[30] Deligkiozi I, Papadakis R, Tsolomitis A. Synthesis, characterisation and photoswitchability of a new [2] rotaxane of α-cyclodextrin with a diazobenzene containing π-conjugated molecular dumbbell. Supramolecular Chemistry. 2012;**24**:333-343. DOI: 10.1080/10610278.2012.660529

[31] Qu D-H, Wang Q-C, Tian H. A half adder based on a photochemically driven [2] rotaxane. Angewandte Chemie, International Edition. 2005;**44**:5296-5299

[32] Sauvage J-P, Dietrich-Buchecker C, editors. Molecular Catenanes, Rotaxanes and Knots: A Journey Through the World of Molecular Topology. 1st ed. Weinheim: Wiley VCH; 1999. ISBN: 3527295720

[33] Rau H. Photo- und Halbleitfähigkeit von Festkörpern isomerer und tautomerer Moleküle. II. Photoleitfähigkeit von 1-Benzolazo-2-naphthol. Berichte der Bunsen-Gesellschaft für Physikalische Chemie. 1969;**73**:810-819. DOI: 10.1002/bbpc.19690730814

[34] Champ RB, Shattuck MD. U.S. Patent 3,898,084. 1975

[35] Deligkiozi I, Papadakis R, Tsolomitis A. Photoconductive properties of a π-conjugated α-cyclodextrin containing [2]rotaxane and its corresponding molecular dumbbell. Physical Chemistry Chemical Physics. 2013;**15**:3497-3503. DOI: 10.1039/C3CP43794A

[36] Deligkiozi I, Voyiatzis E, Tsolomitis A, Papadakis R. Synthesis and characterization of new azobenzene-containing bis pentacyanoferrate(II) stoppered push-pull [2]rotaxanes, with α- and β-cyclodextrin. Towards highly medium responsive dyes. Dyes Pigment. 2015;**113**:709-722. DOI: 10.1016/j.dyepig.2014.10.005

[37] Cacialli F, Wilson JS, Michels JJ, Daniel C, Silva C, Friend RH, Severin N, Samorì P, Rabe JP, O'Connell MJ, Taylor PN, Anderson HL. Cyclodextrin-threaded conjugated polyrotaxanes as insulated molecular wires with reduced interstrand interactions. Nature Materials. 2002;**1**:160-164. DOI: 10.1038/nmat750

[38] Haque SA, Park JS, Srinivasarao M, Durrant JR. Molecular-level insulation: An approach to controlling interfacial charge transfer. Advanced Materials. 2004;**16**:1177-1181. DOI: 10.1002/adma.200400327

[39] Clifford JN, Martínez-Ferrero E, Viterisi A, Palomares E. Sensitizer molecular structure-device efficiency relationship in dye sensitized solar cells. Chemical Society Reviews. 2011;**40**:1635-1646. DOI: 10.1039/b920664g

[40] Monk PMS. The Viologens: Physicochemical Properties, Synthesis and Applications of the Salts of 4,40-Bipyridine. 1st ed. Chichester: John Wiley & Sons Ltd; 1999. ISBN: 978-0-471-98603-4

[41] Huo P, Xue LJ, Li YH, Chen T, Yu L, Zhu QY, Dai J. Effects of alkyl chain length on film morphologies and photocurrent responses of tetrathiafulvalenebipyridinium charge-transfer salts: A study in terms of structures. CrystEngComm. 2016;**18**:2894-2900. DOI: 10.1039/c5ce02479j

[42] Striepe L, Baumgartner T. Viologens and their application as functional materials. Chemistry, a European Journal. 2017;**23**(67):16924-16940. DOI: 10.1002/chem.201703348

[43] Papadakis R, Deligkiozi I, Giorgi M, Faure B, Tsolomitis A. Supramolecular complexes involving non-symmetric viologen cations and hexacyanoferrate (II) anions. A spectroscopic, crystallographic and computational study. RSC Advances. 2016;**6**:575-585. DOI: 10.1039/C5RA16732A

[44] Kinuta T, Sato T, Tajima N, Kuroda R, Matsubara Y, Imai Y. Solid-state thermochromism observed in charge-transfer complex composed of binaphthol and viologen. Journal of Molecular Structure. 2010;**982**:45-49. DOI: 10.1016/j.molstruc.2010.07.048

[45] Raymo FM, Stoddart JF. Interlocked macromolecules. Chemical Reviews. 1999;**99**:1643-1664. DOI: 10.1021/cr970081q

[46] Sheeney-Haj-Ichia L, Willner I. Enhanced photoelectrochemistry in supramolecular CdS-Nanoparticle-stoppered pseudorotaxane monolayers assembled on electrodes. The Journal of Physical Chemistry. B. 2002;**106**:13094-13097. DOI: 10.1021/jp022102c

[47] Feng M, Guo X, Lin X, He X, Ji W, Du S, Zhang D, Zhu D, Gao H. Stable, reproducible nanorecording on rotaxane thin films. Journal of the American Chemical Society. 2005;**127**: 15338-15339. DOI: 10.1021/ja054836j

[48] Saha S, Flood AH, Stoddart JF, Impellizzeri S, Silvi S, Venturi M, Credi A. A redox-driven multicomponent molecular shuttle. Journal of the American Chemical Society. 2007;**129**: 12159-12171. DOI: 10.1021/ja0724590

[49] Sprenger HE, Ziegenbein W. Condensation products of squaric acid and tertiary aromatic amines. Angewandte Chemie, International Edition in English. 1966;**5**:894-894. DOI: 10.1002/anie.196608941

[50] Gassensmith JJ, Baumes JM, Smith BD. Discovery and early development of squaraine rotaxanes. Chemical Communications. 2009:6329-6338. DOI: 10.1039/b911064j

[51] Champ RB, Shattuck MD. U.S. Patent 3,824,099. 1974

[52] Kampfer H. U.S. Patent 3,617,270. 1971

[53] Yan H, Choe HS, Nam SW, Hu Y, Das S, Klemic JF, Ellenbogen JC, Lieber CM. Programmable nanowire circuits for nanoprocessors. Nature. 2011;**470**:240-244. DOI: 10.1038/nature09749

[54] Yum JH, Walter P, Huber S, Rentsch D, Geiger T, Nüesch F, De Angelis F, Grätzel M, Nazeeruddin MK. Efficient far red sensitization of nanocrystalline TiO_2 films by an unsymmetrical squaraine dye. Journal of the American Chemical Society. 2007;**129**:10320-10321. DOI: 10.1021/ja0731470

[55] Leigh DA, Murphy A, Smart JP, Slawai AMZ. Glycylglycine rotaxanes—The hydrogen bond directed assembly of synthetic peptide rotaxanes. Angewandte Chemie, International Edition. 1997;**36**:728-732. DOI: 10.1002/anie.199707281

[56] Gatti FG, Leigh DA, Nepogodiev SA, Slawin AMZ, Teat SJ, Wong JKY. Stiff and sticky in the right places: The dramatic influence of preorganizing guest binding sites on the hydrogen bond-directed assembly of rotaxanes. Journal of the American Chemical Society. 2001;**123**:5983-5989. DOI: 10.1021/ja001697r

[57] Kuhn H. A quantum-mechanical theory of light absorption of organic dyes and similar compounds. The Journal of Chemical Physics. 1949;**17**:1198. DOI: 10.1063/1.1747143

[58] Frampton MJ, Anderson HL. Insulated molecular wires. Angewandte Chemie, International Edition. 2007;**46**:1028-1064. DOI: 10.1002/anie.200601780

[59] Terao J, Tanaka Y, Tsuda S, Kambe N, Taniguchi M, Kawai T, Saeki A, Seki S. Insulated molecular wire with highly conductive π-conjugated polymer core. Journal of the American Chemical Society. 2009;**131**:18046-18047. DOI: 10.1021/ja908783f

[60] Kostromin SV, Malov VV, Tameev AR, Bronnikov SV, Farcas A. The photovoltaic effect and charge carrier mobility in layered compositions of bithiophene or related rotaxane copolymer with C70 fullerene derivative. Technical Physics Letters. 2017;**43**:173-176. DOI: 10.1134/S1063785017020079

[61] Terao J, Wadahama A, Matono A, Tada T, Watanabe S, Seki S, Fujihara T, Tsuji Y. Design principle for increasing charge mobility of p-conjugated polymers using regularly localized molecular orbitals. Nature Communications. 2013;**4**:1691. DOI: 10.1038/ncomms2707

[62] Terao J. Permethylated cyclodextrin-based insulated molecular wires. Polymer Chemistry. 2011;**2**:2444-2452. DOI: 10.1039/c1py00243k

[63] Simone DL. The Synthesis and Investigation of the Electronic Properties of Crown Ether, [2]-Catenane, and [2]-Rotaxane Architectures. Cambridge, Massachusetts: Massachusetts Institute of Technology; 2002

[64] Simone DL, Swager TM. A conducting poly(cyclophane) and its poly([2]-catenane). Journal of the American Chemical Society. 2000;**122**:9300-9301. DOI: 10.1021/ja000970m

[65] Shi J, Xu Y, Wang X, Zhang L, Zhu J, Pang T, Bao X. Synthesis and evaluation of a novel rhodamine B pyrene [2]rotaxane as an intracellular delivery agent for doxorubicin. Organic & Biomolecular Chemistry. 2015;**13**:7517-7529. DOI: 10.1039/C5OB00934K

[66] Barat R, Legigan T, Tranoy-Opalinski I, Renoux B, Péraudeau E, Clarhaut J, Poinot P, Fernandes AE, Aucagne V, Leigh DA, Papot S. A mechanically interlocked molecular system programmed for the delivery of an anticancer drug. Chemical Science. 2015;**6**: 2608-2613. DOI: 10.1039/C5SC00648A

[67] Yu G, Wu D, Li Y, Zhang Z, Shao L, Zhou J, Hu Q, Tang G, Huang F. A pillar[5]arene-based [2]rotaxane lights up mitochondria. Chemical Science. 2016;**7**:3017-3024. DOI: 10.1039/C6SC00036C

[68] Fredy JW, Scelle J, Ramniceanu G, Doan BT, Bonnet CS, Tóth É, Ménand M, Sollogoub M, Vives G, Hasenknopf B. Mechanostereoselective one-pot synthesis of functionalized head-to-head cyclodextrin [3]rotaxanes and their application as magnetic resonance imaging contrast agents. Organic Letters. 2017;**19**:1136-1139. DOI: 10.1021/acs.orglett.7b00153

[69] Erbas-Cakmak S, Leigh DA, McTernan CT, Nussbaumer AL. Artificial molecular machines. Chemical Reviews. 2015;**115**:10081-10206. DOI: 10.1021/acs.chemrev.5b00146

[70] Feringa BL. In control of motion: From molecular switches to molecular motors. Accounts of Chemical Research. 2001;**34**:504-513. DOI: 10.1021/ar0001721

[71] Kottas GS, Clarke LI, Horinek D, Michl J. Artificial molecular rotors. Chemical Reviews. 2005;**105**:1281-1376. DOI: 10.1021/cr0300993

[72] Xue M, Yang Y, Chi X, Yan X, Huang F. Development of pseudorotaxanes and rotaxanes: From synthesis to stimuli-responsive motions to applications. Chemical Reviews. 2015; **115**:7398-7501. DOI: 10.1021/cr5005869

Chapter 3

Ion-Beam Modified Terahertz GaAs Photoconductive Antenna

Caiming Sun and Aidong Zhang

Additional information is available at the end of the chapter

http://dx.doi.org/10.5772/intechopen.79693

Abstract

Ion-implanted photoconductive GaAs terahertz (THz) antennas were demonstrated to deliver both high-efficiency and high-power THz emitters, which are attributed to excellent carrier acceleration and fast carrier trapping for THz generations by analyzing ultrafast carrier dynamics at subpicosecond scale. The implantation distance at over 2.5 μm is deep enough to make defects (Ga vacancies, $As_{Ga}^{+}\ldots$, etc.) quite few; hence, a few with good mobility similar to bare GaAs ensures excellent carrier acceleration in shallow distance <1.0 μm as photo carriers are generated by the pump laser. The implantation dosage is carefully optimized to make carrier trapping very fast, and screen effects by photo-generated carriers are significantly suppressed, which increases the THz radiation power of SI-GaAs antennas by two orders of magnitude. Under the same photo-excitation conditions (pump laser power, bias voltage), photocurrents from GaAs antennas with optimum conditions 300 keV, 5×10^{14} cm^{-2} for H implantation are decreased by two orders of magnitude; meanwhile, the THz radiation is enhanced by over four times, which means that the electrical-to-THz power conversion efficiency is improved by a factor of over 1600.

Keywords: terahertz antenna, ion-beam technology, electrical-to-THz power conversion efficiency, photoconductive antenna, semi-insulating GaAs, ion-implantation

1. Introduction

The development of a terahertz (THz) source has obtained much interest over the last three decades due to their widespread scientific and military applications [1–3]. Photoconductive antennas (PCAs) illuminated by a femtosecond (fs) laser have been becoming the dominant methods for intense THz radiations [3] since the pioneering demonstration of picosecond photoconducting Hertzian dipoles in 1984 [4]. Historically, commercial semi-insulating (SI)-GaAs grown by liquid-encapsulated Czochralski has been the cost-effective choice as the substrate for PCAs, due to its high resistivity (>10^7 Ω cm) and high electron mobility ($\mu > 7000$ cm^2/Vs) [5]. Afterward, about 1 μm-thick film of low-temperature (LT) grown GaAs (LT-GaAs) by molecular beam epitaxy (MBE) on the surface of SI-GaAs substrate is extensively used to reduce carrier lifetime to below 1 ps with high resistivity (10^7 Ω cm) and relatively reasonable mobility μ (100–300 cm^2/Vs) [6, 7], in order to efficiently generate broadband THz radiations of over 1 THz and reduce the carrier lifetime of PCAs on SI-GaAs ($\tau > 100$ ps). An alternative approach for short lifetime is to create point defects in SI-GaAs by ion-implantation technique. Arsenic, oxygen, nitrogen, carbon, hydrogen (proton), etc., have been implanted into SI-GaAs and the obtained GaAs PCAs are similar to those on LT-GaAs [8–11]. However, the process conditions for either LT-GaAs or ion-implanted GaAs are not easy to reproduce in mass production, because of the difficult control of low-temperature process for MBE [12, 13], extremely high implantation energies (~MeV) for heavy ions [11] and the challenging control for post annealing at relatively low temperatures [8, 14].

Fundamentally, the THz radiation power and optical-to-THz conversion efficiency for GaAs PCAs illuminated by femtosecond laser pulses are proportional to the photoconductive material factor $\mu\tau^2$ of PCAs [15]. The reduced electron mobility and carrier lifetime as mentioned earlier will seriously affect the THz power and conversion efficiency [3, 16, 17]. The radiation mechanism is attributed to a time-varying current, a result of photo carriers accelerating across photoconductive gap in the presence of applied electrical field [18]. The emitted THz pulse energy is derived from that stored in the static bias field [19]. Grischkowsky has reported that an extremely strong field exists near the anode of electrically biased PCAs [20], and Salem also demonstrated that THz amplitude can be multiplied by many times when the focused laser beam moved to the anode of an ion-implanted GaAs antenna at the same bias voltage [9]. Recently, plasmonic contact electrodes were used to enhance light absorption within distances of ~100 nm from the anode and 7.5% optical-to-THz conversion efficiency was recorded at very low pump densities of <10 μJ/cm^2 [21]. However, a tightly focused laser beam will cause a high screening effect [22] and the THz power from PCA becomes lower and lower as optical pump saturates [23, 24], which principally sets an upper limit for the conversion efficiency of THz radiations. Thus, it is critical to find out a strategy of creating sufficient defects to reduce the carrier lifetime without affecting mobility detrimentally. High-energy and low-dosage ion-implantation has been verified to be an efficient method of creating proper profiles of defects, in order to obtain both excellent carrier acceleration at the shallow region and fast carrier trapping at the deep layer for THz generations [11, 25]. Also, hydrogen implantation is extensively used to separate high-power active devices (IGBT, laser diodes, LED, etc.) from their mother substrate and get superior performance of high frequency and high efficiency.

2. Experimental results and discussions

In this work, the photoconductive antenna substrate was a commercial high-resistivity (5×10^7 Ω cm), liquid-encapsulated Czochralski-grown, (100)-oriented, semi-insulating (SI)-GaAs wafer. A 10/200 nm-thick Ti/Au metal layer stack was deposited on GaAs substrate by e-beam evaporation, functioning as metal electrodes for PCAs. The PCA has a bow-tie antenna structure with a photoconductive gap of 0.4 mm, 90° bow angle, and antenna length of 2 mm, as shown in **Figure 1(a)**. Afterward, ion beam for hydrogen, helium, or oxygen was implanted into such SI GaAs PCAs with penetration depth of 2.5 μm and the acceleration energies are 300 keV, 800 keV, 3 MeV for H, He, O respectively. Implantation energies were selected so that the peak of ion concentration profile is situated deeper than the thickness of THz-relevant layer within SI-GaAs antenna, which is ~2 μm. The implantation dosage varied from 1×10^{12} to 1×10^{15} cm^{-2}, where lower dose and higher energies were used for heavier ions whereas higher dose and lower energies were suitable for lighter ions. The process details for all samples with different implantation conditions are shown in **Table 1**.

As discussed in our previous work [25], 300 keV H implantation was an easily reproducible condition for fabricating SI-GaAs PCAs with ion penetration depth of 2.5 μm, effectively defining the active region for THz generations. The implantation dosage of H ions varied from 1×10^{14} to 1×10^{15} cm^{-2}, to find out optimum conditions for ion-implanted GaAs PCAs. Density profiles of the generated defects and the implanted ions were optimized by the stopping and range of ions in matter (SRIM) program [26], and the peak distribution situates as deep as 2.5 μm distance from PCA surface. The defects concentration in shallow regions (<1 μm deep) are over three orders of magnitude lower than the peak concentration at 2.5 μm distance, where most photo carriers are generated under femtosecond laser excitation and accelerated at local electrical fields for THz generations. As carriers transit into defects-rich regions underneath the acceleration layer, they will be efficiently trapped and the carrier

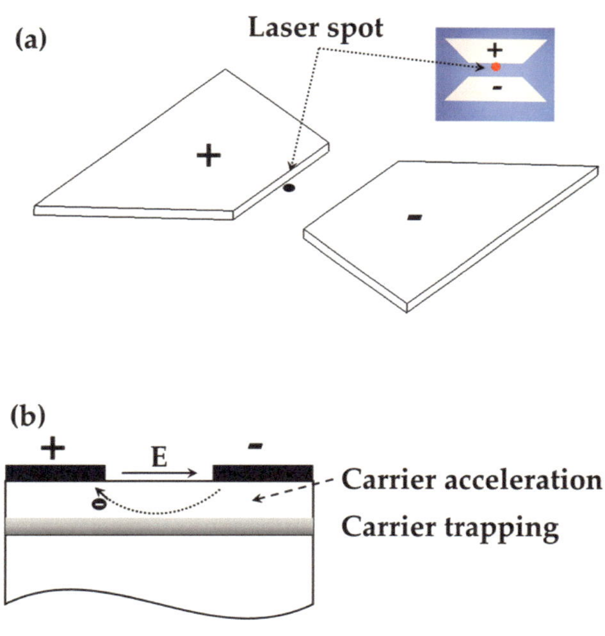

Figure 1. (a) Schematic structure for bow-tie photoconductive antenna (the inset is its photograph). (b) Cross-section of H-implanted GaAs PCAs with acceleration and trapping of photo carriers.

Implantation dose (cm^{-2})	Implantation energy	Ion beam
1×10^{15}	300 KeV	Hydrogen
5×10^{14}	300 KeV	Hydrogen
1×10^{14}	300 KeV	Hydrogen
1×10^{13}	800 KeV	Helium
5×10^{12}	800 KeV	Helium
5×10^{12}	3 MeV	Oxygen
1×10^{12}	3 MeV	Oxygen

Table 1. Process details for all samples with different implantation conditions.

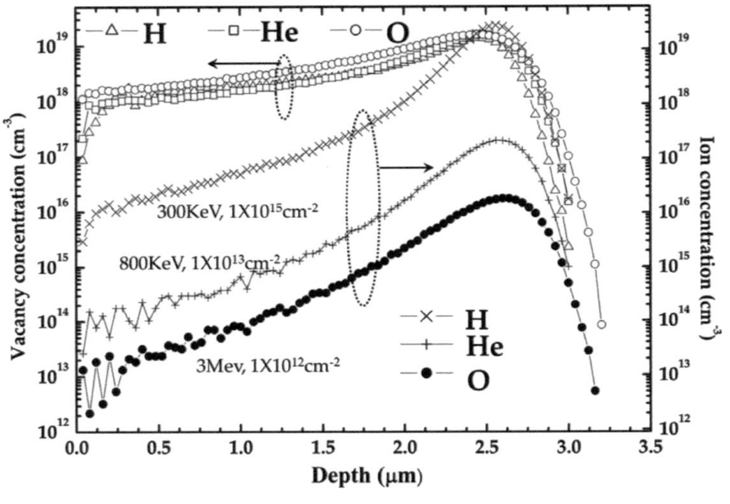

Figure 2. Ion implantation and corresponding vacancy profiles for samples with H dosage of 1×10^{15} cm^{-2}, He dosage of 1×10^{13} cm^{-2}, and O dosage of 1×10^{12} cm^{-2}, calculated by Stopping and Range of Ions in Matter (SRIM), where a Monte-Carlo simulation of 5×10^5 ions was performed for hydrogen, helium, and oxygen. Also, the implantation angle of 7° was used to avoid possible channeling to the crystal axis.

acceleration for THz generations is successfully separated from defects by implantation without obvious decrease of transient mobility in shallow regions. Moreover, the accumulation of photo carriers against the electrical bias was significantly suppressed, avoiding the screen effects by the pump laser (**Figure 1(b)**). Two more sets of samples fabricated on a bare SI-GaAs substrate and 1-μm-thick LT-GaAs grown on an SI-GaAs substrate ("LT-GaAs" hereafter) were prepared for reference.

As shown in **Figure 2**, density profiles of the generated defects and the implanted ions were simulated with the stopping and range of ions in matter (SRIM) program [26]. The implantation conditions for H, He and O are 300 keV, 1×10^{15} cm^{-2}; 800 keV, 1×10^{13} cm^{-2}; and 3 MeV,

1×10^{12} cm^{-2} respectively. At such implantation energies, most ions for all three kinds of samples are implanted far below a surface layer of about 2.5 μm deep where most laser power is absorbed within 1 μm-deep distance and most terahertz power is generated within 2 μm-deep distance. The ion concentration in this region is over three orders of magnitude lower than the peak concentration, whereas the vacancy density profile for all ion beams (H, He, or O) is nearly the same within THz generation distance (~2 μm below the surface). Another reason to use such implantation energies is to reduce the lifetime and therefore strongly reduce the density of photo carriers produced at a depth of ~2 μm, which are not only useless for terahertz generation but also cause the saturation for optical pump.

Figure 3 shows the dark currents versus bias voltage (I-V) characteristics for H-implanted and SI-GaAs PCAs without light illumination, measured by a Keysight B1500A semiconductor device analyzer. Currents passing through the SI-GaAs sample quickly exceed the space-charge limited (SCL) electron flows at low voltage of ~20 V, and then significantly increase by a polynomial V^3 response dominated by a double carrier injection current, as demonstrated in Ref. [21]. It is observed that currents in SI-GaAs antenna under 140 V go up to 10 mA even without any light illumination. This means a considerable temperature increase in active region due to huge heat dissipations, which would affect the efficiency of SI-GaAs antenna and result in electrical breakdown of the device. On the contrary, H-implanted samples follow almost linear dependence of currents on bias voltages in broad range of over 100 V and did not show obvious currents increase of V^3 response even as the bias voltages go up over 200 V. It can be seen that H ions in GaAs extend the linear range of currents dependence on voltages, compared to that for bare SI-GaAs samples. The carrier accumulation near the high electrical field region along anode is significantly suppressed by the corresponding trapping sites and the double carrier injection is eliminated effectively.

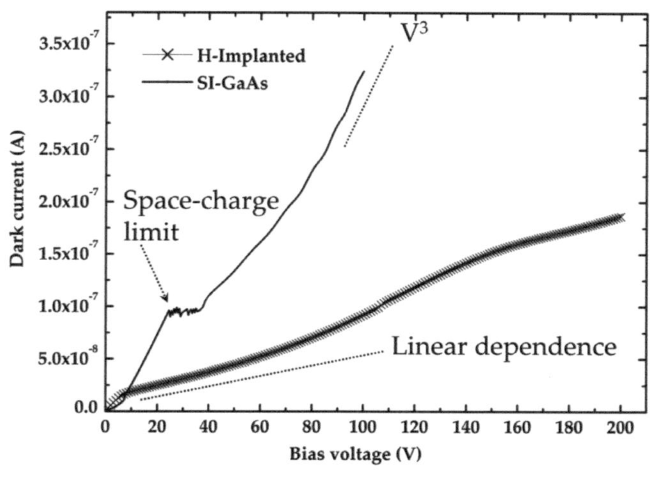

Figure 3. Current-voltage characteristics for SI-GaAs THz antenna and H-implanted one.

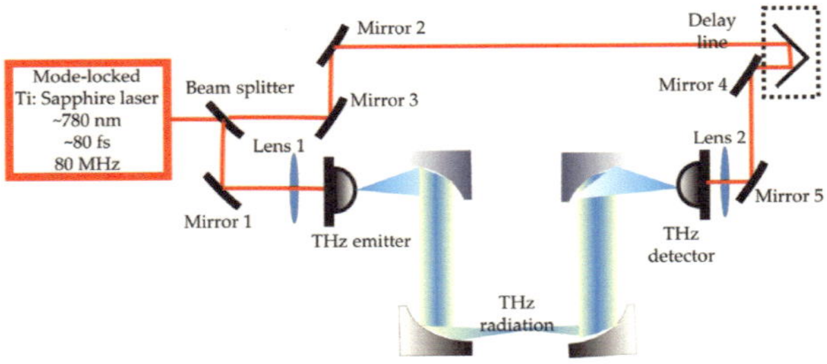

Figure 4. Schematic diagram of THz TDS setup used for measuring electric field of THz pulse.

The setup for characterization of the THz waves is based on a conventional time-domain spectroscopy system (TDS) triggered by a femtosecond laser as shown in **Figure 4**. A mode-locked Ti: Sapphire laser generates 80 fs light pulses at a wavelength of ~780 nm and a repetition rate of 80 MHz. The femtosecond pump pulses were focused by an objective lens with 10 μm-diameter illumination spot on the proximity of anode for a biased photoconductive antenna, which was mounted on the flat side of a Si hemispherical lens with a diameter of 15 mm. The emitted THz radiation was collimated and focused by two pairs of gold-coated off-axis parabolic mirrors onto a photoconductive sampling detector, which was also a photoconductive antenna with bow-tie shape and gap of 20 μm mounted on the back of a Si hemispherical lens with the same diameter. The photoconductive detector was gated by femtosecond probe beam pulses that were separated from the pump beam pulses by a beam splitter.

As_{Ga}^{+}**Figure 5(a)** shows the THz waveforms emitted from GaAs PCAs implanted with H dose of 1×10^{14}, 5×10^{14}, and 1×10^{15} cm^{-2}. The applied bias voltage and pumping power were about 260 V and 60 mW, respectively. The SI-GaAs and LT-GaAs samples were measured under the pump power of 60 mW for reference and the bias voltages for SI-GaAs (40 V) and LT-GaAs (100 V) were kept below breakdown voltages. The waveforms for all samples are normalized at the main peak amplitude in order to see their emission mechanisms clearly. We observe the most sharp THz pulse from GaAs antenna implanted with dose of 5×10^{14} cm^{-2}, and its full width at half maximum (FWHM) of the main peak amplitude is as narrow as 0.3 ps (solid curve), which indicates short carrier lifetimes in the ion-implanted THz emitter and ~0.85 ps lifetime is confirmed by the pump-probe reflectance measurement. The other H-implanted PCAs also demonstrate very short THz pulses with FWHM of 0.31 ps for dose of 1×10^{14} cm^{-2} (dotted curve) and 0.33 ps for dose of 1×10^{15} cm^{-2} (dashed curve). The minimum peak after the main peak for the 5×10^{14} cm^{-2} antenna is sharper than that for all other GaAs PCAs, and its minimum peak before the main peak is also most sharp. The sharp trends of THz signal increase before the maximum THz pulse peak for samples H-1×10^{14} and H-5×10^{14} cm^{-2}, accompanying the current surge in photoconductive region, are completely identical to that case for SI-GaAs sample, which indicates that the carrier mobility in the shallow surface layer for H-1×10^{14} and H-5×10^{14} cm^{-2} PCAs is very close to the mobility of bare SI-GaAs materials. The quite few and uniform point defects (Ga vacancies, As_{Ga}^{+}, etc.) in shallow layer contribute the excellent mobility of these H-implanted samples, compared to LT-GaAs and the sample of H-1×10^{15} cm^{-2}. In order to further identify the uniqueness of these

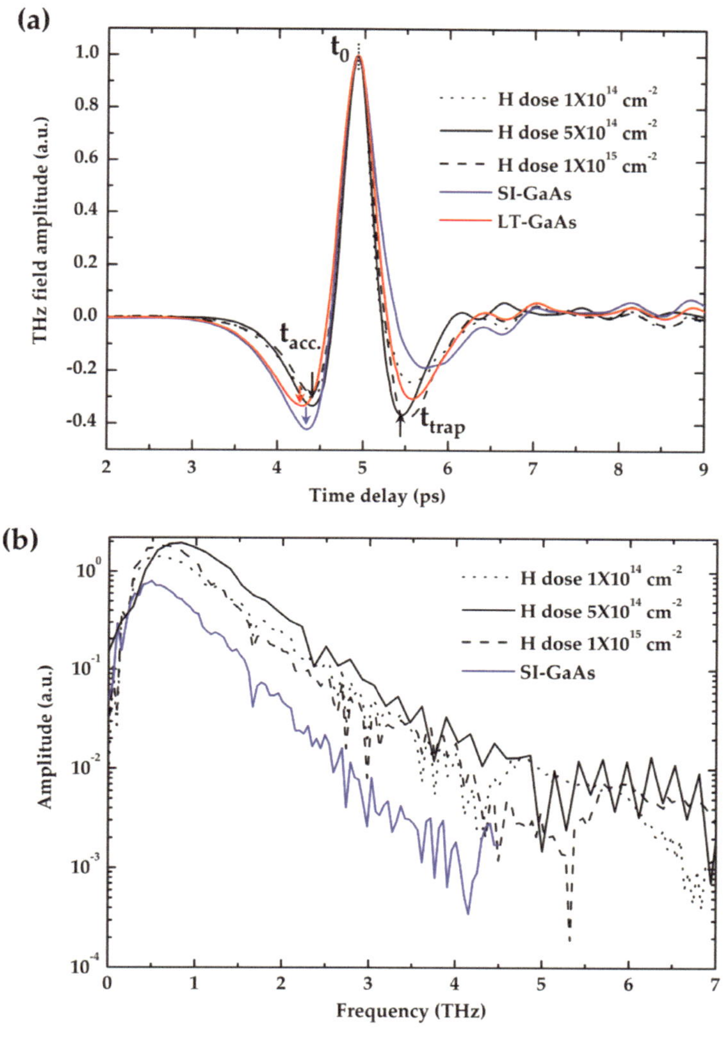

Figure 5. (a) Normalized THz pulses emitted from GaAs antennas of 1×10^{14}, 5×10^{14}, 1×10^{15} cm^{-2}, SI-GaAs, and LT-GaAs. (b) Fast Fourier transformed spectra of THz signals for different GaAs emitters.

ion-implanted GaAs antennas, the fast Fourier transformed (FFT) spectra of the waveform for all implanted PCAs and the reference samples of SI-GaAs are shown in **Figure 5(b)**. We find that ion-implanted GaAs antennas generate THz signals with frequency of over 5 THz and the sample with dose of 5×10^{14} cm^{-2} demonstrates the strongest signals in the high-frequency range of 1 –5 THz. As expected, SI-GaAs samples produce the weakest signals at high frequencies of over 1 THz among all samples.

In order to interpret the THz radiation waveform, we analyze the THz radiation assuming that the emitted field $E_{THz}(t)$ is proportional to the time derivative of the transient current $J(t)$ at far field, as expressed in Eq. (1).

$$E_{THz}(t) \propto \int \frac{\partial J(t)}{\partial t} dxdydz \tag{1}$$

where the integration is taken over the whole device including carrier acceleration layer and carrier trapping layer (**Figure 1(b)**). The transient current depends on the free-carrier concentration n and on the mean velocity v of the electrons:

$$J(t) = -en(t)v(t) \tag{2}$$

the contribution of the holes which have a much smaller mobility is neglected. We analyze the carrier transport based on a set of kinetic equations [27] which can be written as follows:

$$\frac{dn(t)}{dt} = -\frac{n(t)}{\tau_c} + G(t) \tag{3}$$

$$\frac{dv(t)}{dt} = -\frac{v(t)}{\tau_m} + \frac{eE_{loc}}{m^*} - \frac{v(t)G(t)}{n(t)} \tag{4}$$

where G is a photoinjection rate, E_{loc} is the local electric field, τ_c is the free electron lifetime, and τ_m is the momentum relaxation time, which is connected to the mobility of the free electrons in shallow layer.

Based on this theoretical model, the main positive peak observed in the waveforms of **Figure 5 (a)** is attributed to the rises of surge current by photo-carrier injection and the subsequent carrier acceleration under bias fields ($t_{acc.}$ in **Figure 5(a)**), while the second negative peak after the main peak is related to the decay of current governed by the carrier trapping (t_{trap} in **Figure 5(a)**). For the ion-implanted GaAs PCAs, we have to consider the carrier dynamics of acceleration process in shallow layer (<2 μm) and trapping process in the underneath layer (~2.5 μm deep), as shown in **Figure 1(b)**. After laser is absorbed within 1 μm depth, photo carriers are created accordingly and accelerated within 2 μm depth for efficient THz generations. The main peak distribution of implanted ions and related defects at ~2.5 μm depth enables efficient carrier trapping and significantly reduce carrier concentration in the trapping layer (**Figure 1(b)**). If the pump pulse laser width (δt) is larger than the carrier momentum relaxation time (τ_m), and if the carrier lifetime (τ_c) is larger than the pump laser pulse width ($\tau_m < \delta t < \tau_c$), the carrier acceleration and resultant current rise are determined by the pump laser pulse width, which is related to $t_{acc.}$. This is exactly the situation in the shallow laser absorption layer (1 μm deep) for ion-implanted GaAs PCAs; where the momentum relaxation time was estimated to be about 10 fs; the laser pulse width was 80 fs, and the carrier lifetime was over 10 ps (similar to lifetime in bare SI GaAs), respectively. Assuming the transition time for photo carriers from absorption region (1 μm) to the trapping layer (2.5 μm) is shorter than the carrier trapping time, the carrier trapping and corresponding current decay depends on peak concentrations of H ions and implantation-related defects at 2.5 μm depth; considering that carrier transition time is about 100 fs, and the carrier trapping time (t_{trap}) is estimated to be ~0.8 ps for the sample H-5 × 10^{14} cm^{-2}. In the progress of current rise dominated by laser pump, carrier acceleration should not be affected by implantation defects because the ion concentration in this region is several orders of magnitude lower than the peak concentration and the carrier mobility is able to keep very close to that in SI GaAs. However, the defects for LT-GaAs and the sample of H-1 × 10^{15} cm^{-2} have decreased the carrier mobility by scattering to some extent that the carrier acceleration turns slow when comparing with SI-GaAs in **Figure 5(a)**, considering the momentum relaxation time (τ_m) may increase to be comparable with the laser pulse width (δt) and the current surge is affected accordingly. Meanwhile, the current decay is dominated by the trapping effect in the underneath THz generation layer. This structure will form vertical confinement for the distribution of photo carriers and block the carriers in SI GaAs substrate, which makes the trapping layer return the original insulating state after the fast carrier trapping. It is noted that the carrier trapping time (t_{trap}) for samples of H-5 × 10^{14} and H-1 × 10^{15} cm^{-2} is significantly shorter than that of H-1 × 10^{14} cm^{-2}, LT-GaAs, and SI-GaAs samples because the latter did not have efficient structures for vertical confinement of photo carriers. Therefore, we conclude that this confinement structure for photoconductive antennas will relieve the screen effect caused by charges accumulation in photoconductive region and reduce the saturation effect by laser excitation.

In **Figure 6(a)**, we show variation of the peak of emitted THz field amplitude with the pump laser power for all samples measured under TDS in **Figure 4**. The bias voltage of 140 V was used. The SI-GaAs sample without H implantation became saturated as the pump laser power exceeded 30 mW, similar to the reports in Refs. [6–10]. Thermal breakdown of SI-GaAs emitters easily occurred as they are saturated by the pump laser and the bias voltage. Normally, SI-GaAs emitters are recommended to operate far enough away from the saturation status. Hydrogen-implanted GaAs emitter with dose of 1 × 10^{14} cm^{-2} showed relatively linear increase of THz amplitude as the pump laser power. We are able to get the maximum THz field 3.5 times bigger than that from SI-GaAs emitter but no obvious saturation is found at the laser power of over 60 mW. The H-5 × 10^{14} cm^{-2} sample provides the best performance that almost linear dependence of THz fields emitted on the pump laser power is demonstrated, and the maximum THz field we could obtain from H-5 × 10^{14} cm^{-2} sample is five times bigger than that from SI-GaAs emitter. It should be emphasized that the H-implanted GaAs with dose of 5 × 10^{14} cm^{-2} did not show any saturation property with the pump laser approaching 100 mW in the TDS measurement range and the bias voltage increasing up to 260 V. The sample of H-1 × 10^{15} cm^{-2} showed deteriorated mobility, and THz fields are smaller than those emitted from sample 1 × 10^{14} cm^{-2}. Because point defects created by H implantation at lower dose of 1 × 10^{14}–5 × 10^{14} cm^{-2} are more uniform and fewer, carrier mobility is kept to be very close to that of SI-GaAs with the carrier momentum relaxation time (τ_m) as short as 10 fs also. In the optimum operation conditions (i.e., 80 mW of the laser power and 260 V of the bias voltage for 5 × 10^{14} cm^{-2} sample versus 30 mW of the laser power and 140 V of the bias voltage for SI-GaAs), THz power emitted from the H-implanted sample was ~100 times bigger than that of traditional SI-GaAs THz emitter.

Figure 6(b) presents the photocurrent as a function of the bias voltage (*I-V*) at a pump laser power of 60 mW for all H-implanted and SI-GaAs samples. Fairly high photocurrents and saturation behavior are observed for the SI-GaAs sample. With increasing the dosage of H implantation, there is continuous decrease in the photocurrents for the ion-implanted GaAs PCAs compared to the conventional SI-GaAs devices, because the presence of a deep trapping layer at 2.5 μm depth blocks the photo carriers in the SI-GaAs substrate. This means that heat

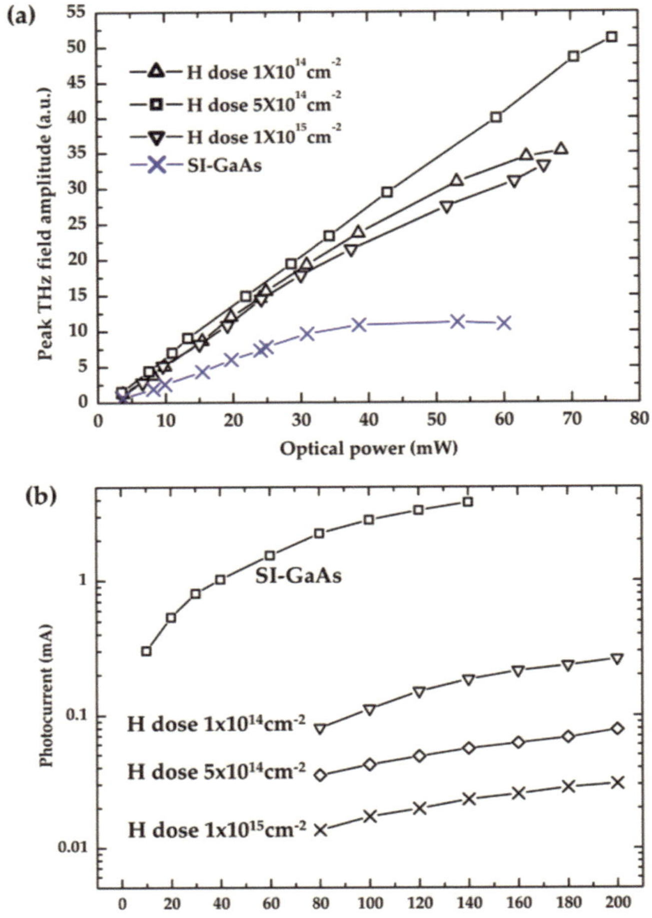

Figure 6. (a) Peak THz field amplitude from GaAs emitters as a function of pump laser power under a bias voltage of 140 V. (b) Photocurrent-voltage (*I-V*) characteristics under excitation at the laser power of 60 mW for H-implanted and SI-GaAs samples. The photocurrent was calculated from the measured current under illumination by subtracting the dark current.

generation will be efficiently suppressed in ion-implanted samples as H dose is increased, then thermal breakdown voltage of GaAs PCAs will become higher as higher dose is utilized. H-implanted GaAs PCAs become electrically robust and are able to stably operate from ~80 to >260 V. To make a rough estimate of improvement in electrical-to-THz power conversion efficiency of the H-implanted GaAs PCAs, we can compare the THz power emission for H-5×10^{14} cm^{-2} sample with conventional SI-GaAs sample without any implantation. As shown in **Figure 6(a)**, the THz field amplitude is about four times more (power will be 16 times) from H-5×10^{14} cm^{-2} sample than SI-GaAs sample under the optical power of 60 mW. Moreover, the corresponding photocurrent for H-5×10^{14} cm^{-2} sample is about 100 times smaller than that for SI-GaAs sample at the same voltage of 140 V. Then, under the same photo-excitation conditions (pump laser power, bias voltage), photocurrent is 100 times smaller, but emitted THz power is 16 times more for H-5×10^{14} cm^{-2} sample than SI-GaAs sample. The electrical to

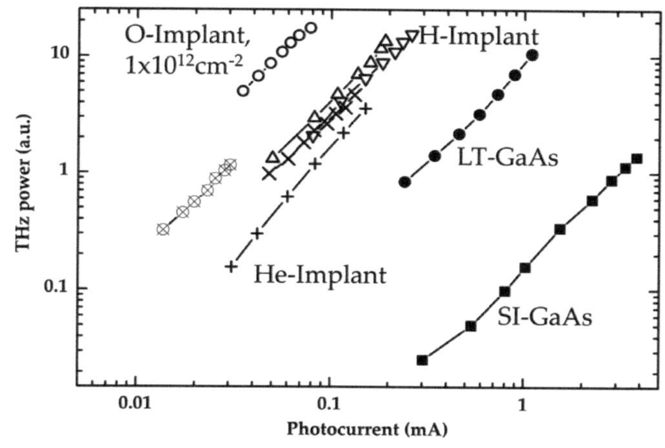

Figure 7. The terahertz power as a function of the photocurrent is represented for ion-implanted GaAs PCAs.

THz power conversion efficiency of H-5 $\times$ 10^{14} cm^{-2} emitter is almost ~1600 times better than SI-GaAs emitter.

To further confirm the mobility in the THz generation layer of ion-implanted PCAs is superior to that of LT-GaAs, **Figure 7** relates the THz radiation power to photocurrents for all ion-implanted PCA samples for a constant laser power in a logarithmic scale. The relationships for all samples are curve-fitted to parallel lines with a slope of 2, indicating the quadratic dependence of the radiation power on the induced photocurrents by the pump laser, and the fact that operation conditions (laser alignment, output THz coupling efficiency, antenna structure, etc.) are the same for all PCAs. The ion-implanted GaAs antennas for H-1 $\times$ 10^{14}, H-1 $\times$ 10^{15}, and O 1 $\times$ 10^{12} cm^{-2} generated stronger THz radiation than LT-GaAs PCAs under the excitation of constant laser power 200 mW, however the DC photocurrents are reduced by about 100 times. This enhancement for THz generation may mainly be attributed to the better quality of photoconductive GaAs with higher mobility than that of LT-GaAs since the conversion efficiency is proportional to the carrier mobility, while the carrier concentration is tightly confined by deep trapping layer at ~2.5 μm-distance in ion-implanted PCAs.

In order to further interpret the THz radiation power and the optical-to-THz conversion efficiency for GaAs PCAs, we deduced the theoretical model accordingly for photoconductive antennas.

According to the Ref. [28], transient photocurrent under fs laser excitations can be written as below Eq. (5).

$$I_{ph} = \frac{eV_b \mu_e \tau \eta_L P_L}{hf_L d^2} \tag{5}$$

where e is the electron charge, f_L is the optical frequency, h is the Planck constant, V_b is the applied bias voltage, μ_e is the electron mobility, τ is the lifetime for photocarriers, η_L is the absorption efficiency for laser illumination, P_L is the incident laser power on the gap, d is the length of photoconductive gap.

Meanwhile, the transient resistance under laser excitations can be approximately formulated as Eq. (6) [15].

$$R_{gap} = \frac{hf_L f_R d^2}{e\mu_e \eta_L P_L} \tag{6}$$

where f_R is the repetition rate for incident fs laser pulses.

Therefore, using Eqs. (5) and (6), we can obtain the expression for the electric power generated between the photoconductive gap, subjected to the pump laser power P_L and the optical-to-THz conversion efficiency for the antenna η_{LE}.

$$P_E = I_{ph}{}^2 R_{gap} = \left(\frac{eV_b \mu_e \tau \eta_L P_L}{hf_L d^2}\right)^2 \frac{hf_L f_R d^2}{e\mu_e \eta_L P_L} = \frac{eV_b{}^2 \mu_e \tau^2 \eta_L P_L f_R}{hf_L d^2} = \frac{eE_b{}^2 \mu_e \tau^2 \eta_L P_L f_R}{hf_L} \tag{7}$$

$$\eta_{LE} = \frac{P_E}{P_L} = \frac{eV_b{}^2 \mu_e \tau^2 \eta_L f_R}{hf_L d^2} = \frac{eE_{loc}{}^2 \mu_e \tau^2 \eta_L f_R}{hf_L} \tag{8}$$

where E_{loc} is the localized electric field around the anode proximity, for separating the electron-hole pairs, accelerating photocarriers, and the generation of transient photocurrents.

As seen from Eq. (8), the optical-to-THz conversion efficiency is directly proportional to the square of the bias voltage V_b and to the photoconductive material factor $\mu\tau^2$.

According to [28] and the references wherein, the saturation behavior of THz radiation amplitude E_r against the pump intensity F_L or pump power P_L can be expressed as

$$E_r \approx \frac{F_L / F_s}{1 + F_L / F_s} = \frac{P_L / P_s}{1 + P_L / P_s} \tag{9}$$

where F_s, P_s are the characteristic saturation intensity and saturation power for the PC antenna respectively.

$$F_s = \frac{hf_L}{e\mu_e \eta_L} \tag{10}$$

Eq. (10) shows that the saturation behavior for PC antennas will easily take place when the carrier mobility of the photoconductive material is high, and F_s is normally below 100 μJ/cm^2 for conventional SI-GaAs and LT-GaAs PCAs [21–24, 29]. Comparably, our GaAs PCAs based on high-energy ion-implantation did not show any saturation property even when the pump laser intensity increases as high as 10 mJ/cm^2, which indicating the carrier mobility and carrier lifetime are quite low in the deep trapping layer. The carrier mobility and carrier lifetime in the shallow layer for THz generation are as high as native SI-GaAs materials, which also means that the photoconductive material factor $\mu\tau^2$ is very high to guarantee the high optical-to-THz conversion efficiency for PCAs.

3. Conclusion

In summary, the GaAs PCAs' saturation effect for the excitation of pump laser is efficiently reduced by hydrogen implantation, due to the vertical confinement of photo carriers in H-implanted emitters. THz emitter implanted by H ions of 300 keV and 5×10^{14} cm^{-2} has both excellent mobility and short enough carrier lifetime. Thus, the optical-to-THz conversion efficiency is improved 16 times and the electrical-to-THz conversion efficiency is 1600 times compared to conventional GaAs emitters. Electrically robust H-implanted GaAs PCA is able to operate from ~80 to >260 V without any thermal breakdown. The emitted THz power from H-implanted GaAs antenna is more than two order of magnitude stronger than that from traditional GaAs emitter.

Acknowledgements

This work was supported by the National Natural Science Foundation of China (Grant No. U1613223). The authors acknowledge Ms. Ho Lai Ching, a staff member at the Department of Electronic Engineering, the Chinese University of Hong Kong, Hong Kong, China, for her assistance in equipment maintenance.

Author details

Caiming Sun[1,2]* and Aidong Zhang[1]

*Address all correspondence to: cmsun@cuhk.edu.cn

1 Institute of Robotics and Intelligent Manufacturing (IRIM), The Chinese University of Hong Kong, Shen Zhen, People's Republic of China

2 Department of Electronic Engineering, The Chinese University of Hong Kong, Shatin, NT, Hong Kong, People's Republic of China

References

[1] Dhillon SS et al. The 2017 terahertz science and technology roadmap. Journal of Physics D: Applied Physics. 2017;**50**(4):043001

[2] Lewis RA. A review of terahertz sources. Journal of Physics D: Applied Physics. 2014;**47**(37):374001

[3] Hafez HA, Chai X, Ibrahim A, Mondal S, Ferachou D, Ropagnol X, Ozaki T. Intense terahertz radiation and their applications. Journal of Optics. 2016;**18**(9):093004

[4] Auston DH, Cheung KP, Smith PR. Picosecond photoconducting Hertzian dipoles. Applied Physics Letters. 1984;**45**(3):284

[5] Makram-Ebeid S, Tuck B. Semi-Insulating III-V Materials. Nantwich: Shiva; 1982

[6] Tani M, Matsuura S, Sakai K, Nakashima S. Emission characteristics of photoconductive antennas based on low-temperature-grown GaAs and semi-insulating GaAs. Applied Optics. 1997;**36**:7853

[7] Tani M, Sakai K, Mimura H. Ultrafast Photoconductive Detectors Based on Semi-Insulating GaAs and InP. Japanese Journal of Applied Physics, Part 2. 1997;**36**:L1175

[8] Liu TA, Tani M, Pan CL. THz radiation emission properties of multienergy arsenic-ion-implanted GaAs and semi-insulating GaAs based photoconductive antennas. Journal of Applied Physics. 2003;**93**(5):2996

[9] Salem B, Morris D, Aimez V, Beerens J, Beauvais J, Houde D. Pulsed photoconductive antenna terahertz sources made on ion-implanted GaAs substrates. Journal of Physics. Condensed Matter. 2005;**17**(46):7327

[10] Salem B, Morris D, Salissou Y, Aimez V, Charlebois S, Chicoine M, Schiettekatte F. Terahertz emission properties of arsenic and oxygen ion-implanted GaAs based photoconductive pulsed sources. Journal of Vacuum Science and Technology A. 2006;**24**(3):774

[11] Singh A, Pal S, Surdi H, Prabhu SS, Mathimalar S, Nanal V, Pillay RG, Döhler GH. Carbon irradiated semi insulating GaAs for photoconductive terahertz pulse detection. Optics Express. 2017;**23**(5):6656

[12] Ludwig C, Kuhl J. Studies of the temporal and spectral shape of terahertz pulses generated from photoconducting switches. Applied Physics Letters. 1996;**69**(9):1194

[13] Kono S, Tani M, Sakai K. Ultrabroadband photoconductive detection: Comparison with free-space electro-optic sampling. Applied Physics Letters. 2001;**79**(7):898

[14] Gregory IS, Baker C, Tribe WR, Evans MJ, Beere HE, Linfield EH, Davies AG, Missous M. High resistivity annealed low-temperature GaAs with 100 fs lifetimes. Applied Physics Letters. 2003;**83**(20):4199

[15] Glinskiya IA, Khabibullin RA, Ponomarev DS. Total Efficiency of the Optical-to-Terahertz Conversion in Photoconductive Antennas Based on LT-GaAs and $In_{0.38}Ga_{0.62}As$. Russian MicroElectronics. 2017;**46**(6):408-413

[16] Huang Y, Khiabani N, Shen Y, Li D. Terahertz photoconductive antenna efficiency. In: Proceedings of the International Workshop Antenna Technology (iWAT); Hong Kong, China; 2011. pp. 152-156

[17] Khiabani N, Huang Y, Shen YC, Boyes SJ. Theoretical modeling of a photoconductive antenna in a terahertz pulsed system. IEEE Transactions on Antennas and Propagation. 2013;**61**(4):1538

[18] Taylor AJ, Benicewicz PK, Young SM. Modeling of femtosecond electromagnetic pulses from large-aperture photoconductors. Optics Letters. 1993;**18**(16):1340

[19] Darrow JT, Zhang X-C, Auston DH. Power scaling of large-aperture photoconducting antennas. Applied Physics Letters. 1991;**58**(1):25

[20] Ralph SE, Grischkowsky D. Trap-enhanced electric fields in semi-insulators: The role of electrical and optical carrier injection. Applied Physics Letters. 1991;**59**(16):1972

[21] Yang S-H, Hashemi MR, Berry CW, Jarrahi M. 7.5% Optical-to-Terahertz Conversion Efficiency Offered by Photoconductive Emitters With Three-Dimensional Plasmonic Contact Electrodes. IEEE Transactions on Terahertz Science and Technology. 2014;**4**(5):575

[22] Kim DS, Citrin DS. Coulomb and radiation screening in photoconductive terahertz sources. Applied Physics Letters. 2006;**88**:161117

[23] Rodriguez G, Caceres SR, Taylor AJ. Modeling of terahertz radiation from biased photoconductors: Transient velocity effects. Optics Letters. 1994;**19**(23):1994

[24] Benicewicz PK, Roberts JP, Taylor AJ. Scaling of terahertz radiation from large-aperture biased photoconductors. Journal of the Optical Society of America B: Optical Physics. 1994;**11**(12):2533

[25] Sun CM, Zhang AD. Efficient terahertz generation from lightly ion-beam-treated semi-insulating GaAs photoconductive antennas. Applied Physics Express. 2017;**10**(10):102202

[26] Ziegler JF, Biersack JP, Littmark U. The Stopping and Range of Ions in Solids. Vol. 1. New York: Pergamon; 1985. See also http://www.srim.org/

[27] Jepsen PU, Jacobsen RH, Keiding RH. Generation and detection of terahertz pulses from biased semiconductor antennas. Journal of the Optical Society of America B: Optical Physics. 1996;**13**(11):2424

[28] Smith PR, Auston DH, Nuss MC. Subpicosecond Photoconducting Dipole Antennas. IEEE Journal of Quantum Electronics. 1988;**24**(2):255-260

[29] Chou R-H, Pan C-L. Gap-Dependent Terahertz Pulses from Mid-Size-Gap Multi-Energy Arsenic-Ion-Implanted GaAs Antennas. Japanese Journal of Applied Physics. 2008;**47**(11): 8419-8425

Chapter 4

Energy Bandgap Engineering of Transmission-Mode AlGaAs/GaAs Photocathode

Yijun Zhang and Gangcheng Jiao

Additional information is available at the end of the chapter

http://dx.doi.org/10.5772/intechopen.80704

Abstract

Aiming to enhance the photoemission capability in the waveband region of interest, a graded bandgap structure was applied to the conventional transmission-mode AlGaAs/GaAs photocathodes based on energy bandgap engineering, wherein the composition in $Al_xGa_{1-x}As$ window layer and the doping concentration in GaAs active layer were gradual. According to Spicer's three-step model, a photoemission theoretical model applicable to the novel transmission-mode $Al_xGa_{1-x}As$/GaAs photocathodes was deduced so as to guide the cathode structural design. Then the cathode material was grown by the metalorganic chemical vapor deposition technique, and the epitaxial cathode material quality was evaluated by the means of scanning electron microscope, electrochemical capacitance-voltage, X-ray diffraction and spectrophotometry. Through a series of specific processes, the cathode material was made into the multilayered module, possessing a glass/Si_3N_4/$Al_xGa_{1-x}As$/GaAs structure. After the surface treatment including heat cleaning and Cs—O activation for the cathode module, the image intensifier tube comprising the activated cathode module, microchannel plate, and phosphor screen was fabricated by indium sealing. The spectral response test results confirm the validity of the novel structure for the enhancement of blue-green photoresponse.

Keywords: AlGaAs/GaAs photocathode, graded bandgap, photoemission model, material epitaxy, tube fabrication

1. Introduction

Since negative-electron-affinity (NEA) GaAs photocathode was proposed as a type of excellent photoemitter by Scheer and Laar [1], GaAs-based photocathodes have found widespread applications in photodetectors, accelerators, electron microscopes, photon-enhanced thermionic emission devices, and other fields [2–5]. In view of the high visible spectral response, good spectral extensibility to the near infrared (NIR) region and low dark current, NEA GaAs, GaAsP, and InGaAs photocathodes are important components in the vacuum photodetectors, for example, low-light-level (LLL) image intensifiers, photomultiplier tubes, and streak tubes [6]. In the modern light sources based on free electron lasers or energy recovery linacs, GaAs-based photocathodes serve as high brightness electron sources with the unique virtues of large current density driven by visible lasers, high spin polarization, low thermal emittance, and narrow energy distribution [7]. In recent years, a spin-polarized transmission electron microscope combining electron microscopy and accelerator technology using GaAs-GaAsP strained superlattice photocathodes was developed to observe dynamically a magnetic field images with high spatial and temporal resolutions [8]. Moreover, with the aid of the ultrahigh speed pulse laser, GaAs photocathodes can satisfy the requirements of fast response speed and large emission current density aiming to THz frequency vacuum devices [9].

As is well known, GaAs photocathodes can operate in the transmission-mode (t-mode) and the reflection-mode (r-mode), respectively, depending on the difference in the direction of the incident light [10, 11], as shown in **Figure 1**. For the t-mode operation, the incident light is irradiated on the substrate surface, and the photoelectrons are extracted from the opposite surface side, whereas for the r-mode operation, the incident light and photoelectrons are located on the same emission surface side. Due to the difference in absorption length of longwave and shortwave photons, the shapes of spectral response curves for GaAs photocathodes working in the two modes are different [11]. Differing from r-mode GaAs photocathodes, t-mode ones are difficult to achieve high spectral response in a broadband region from ultraviolet to NIR spectrum. Usually in the practical applications, the researches on t-mode photocathodes are more concerned. For example, the image intensifiers and related imaging systems, t-mode photocathodes conform to the optical imaging structure [12]. Besides, as polarized electron sources in photoinjector apparatus, t-mode photocathodes

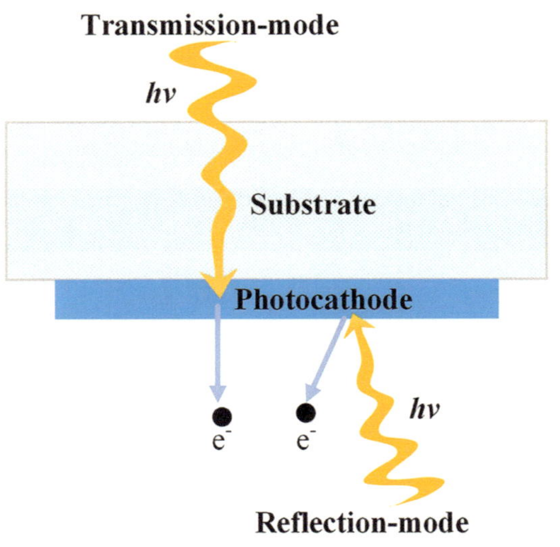

Figure 1. Schematic diagram of thin photocathode operating in the two different modes.

are more popular than r-mode ones, because the laser spot size can be reduced through the short focus lens placed on the photocathode backside, which would not hinder the path of the electron beam and is more conducive to achieve a super-high-brightness electron beam [13, 14].

As proposed by Spicer and Herreragomez [15], the photoemission process from photocathodes consists of electron excitation by incident light absorption, electron transport toward surface, and electron escape across the surface barrier into vacuum. For t-mode photocathodes, some important cathode parameters such as electron diffusion length, interface recombination velocity, and surface escape probability are crucial to the photoemission performance, especially the shortwave photosensitivity [16]. Enhancing the blue-green response of t-mode GaAs photocathodes as far as possible, would not only be beneficial to the detection in sandy or desert terrain for image intensifiers [17], but also increase the current density driven by 532 nm laser for electron sources [18]. Although the external electric field biased across the photocathodes can improve the photoemission capability, the limitations of this approach are the difficulty in making thin electrode pattern and the increased dark current with the strong field [19, 20]. In view of this adverse case, internal built-in electric fields through energy band engineering design could be an alternative approach. In our research, a complex structure composed of the composition-graded structure and the doping-graded structure is proposed to prepare high efficient t-mode AlGaAs/GaAs photocathodes. Furthermore, the photoemission model, cathode structure design, cathode material epitaxy, and vacuum tube fabrication are investigated through the integrated analysis of theory and experiments. Finally, the effectivity of the designed novel structure is verified by comparison with the common photocathodes.

2. Graded bandgap structure

For the t-mode GaAs photocathode, the AlGaAs and GaAs materials are usually used as the window layer and the active layer, which determine the shortwave cutoff and longwave cutoff, respectively. A built-in electric field in the interior of the photocathode material can be realized by the variation of dopant or composition according to energy band engineering design [21, 22]. Based on this concept, a novel structure is proposed to improve photoelectron emission capability, wherein a composition-graded structure and a doping-graded structure are employed to the $Al_xGa_{1-x}As$ window layer and GaAs active layer, respectively [23, 24], as shown in **Figure 2**. To form a built-in constant electrical field in the GaAs active layer of the photocathode, the p-type dopant concentration can follow the exponential variation, and the doping formula is expressed by [22]

$$N(y) = N_0 \exp(Ay) \tag{1}$$

where A denotes the exponential-doping (e-doping) factor, N_0 is the doping concentration at the surface of GaAs active layer, y is the distance from the coordinate origin (i.e., the surface of GaAs active layer), and $N(y)$ is the p-type doping concentration in the GaAs active layer. As a result of the variation of dopant concentration, the initial Fermi level is different. In thermal

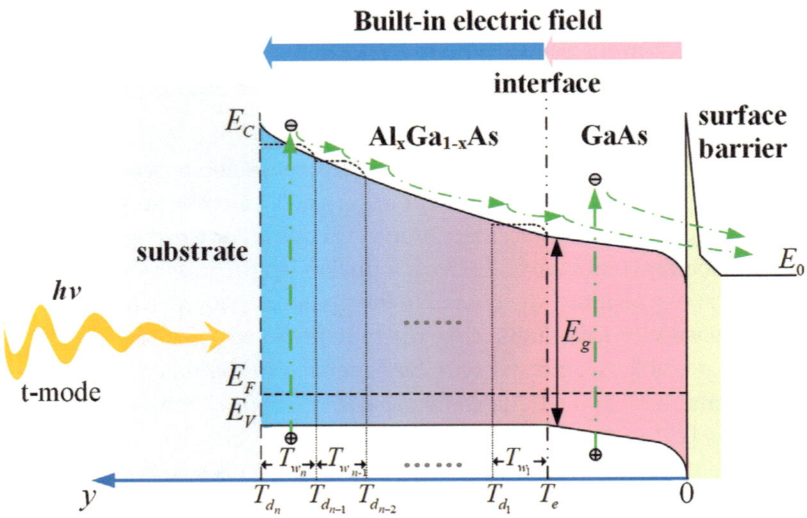

Figure 2. Energy band structure diagram of the t-mode $Al_xGa_{1-x}As/GaAs$ photocathode with the graded bandgap structure. E_g is the bandgap, E_0 is the vacuum level, E_F is the Fermi level, E_C is the conduction band minimum, and E_V is the valence band maximum.

equilibrium, the Fermi level at different positions in the active layer is unified, and the electric potential energy $qV(y)$ is varied as follows:

$$qV(y) = k_B T \ln \frac{N(y)}{N_0} = k_B T A y \tag{2}$$

where q is the electron charge, T is the cathode temperature, and k_B is the Boltzmann constant. The diagram of band structure with the downward shape in the GaAs active layer is shown in **Figure 2**, wherein the built-in electric field E_0 in a certain thick (T_e) active layer is given by

$$E_0 = \frac{dV(y)}{dy} = \frac{k_B T A}{q} = \frac{k_B T}{q T_e} \ln \frac{N(y)|_{y=T_e}}{N_0} \tag{3}$$

In the $Al_xGa_{1-x}As$ window layer, the bandgap is decreased from the substrate interface to the GaAs interface due to the composition-graded structure. Because of the high p-type doping concentration, the valence bands of the $Al_xGa_{1-x}As/GaAs$ heterojunction are aligned, as shown in **Figure 2**. The graded Al composition in the window layer results in a built-in electric field E_1, which is treated to be uniform as follows [25]:

$$E_1 = \frac{\Delta E_g}{q \Delta d} \tag{4}$$

where ΔE_g is the energy bandgap difference of $Al_xGa_{1-x}As$ material, and Δd is the overall thickness of $Al_xGa_{1-x}As$ window layer. **Figure 2** illustrates the transport process of photoexcited electrons in the t-mode graded bandgap $Al_xGa_{1-x}As/GaAs$ photocathodes. As for the t-mode photocathodes, the photoelectrons generated by shortwave light excitation are distributed in the $Al_xGa_{1-x}As$ window layer. Under the first-stage built-in electric field, the thermalized photoelectrons in the $Al_xGa_{1-x}As$ layer move toward the GaAs interface. After that, the second-stage built-in electric field in the active layer can promote these photoelectrons toward the emission surface. On the other hand, the GaAs active layer can absorb the longwave light, and the excitated photoelectrons are promoted to move toward the surface with the help of the built-in electric field in the active layer. Consequently, there are reasons to believe that, by virtue of this unique graded bandgap structure, the quantum efficiency over the broadband spectrum, especially in the shortwave response region would be enhanced to some extent.

3. Photoemission model derivation and simulation

3.1. Photoemission model derivation

As is well known, the one-dimensional continuity equation can afford a useful avenue to establish the photoemission model of t-mode or r-mode III–V group photocathodes, which takes account of the spatial photon adsorption, spatial carrier distribution, and interface electron recombination [10, 11]. As shown in **Figure 2**, the photoelectrons generated in the $Al_xGa_{1-x}As$ layer are able to move into the GaAs layer and contribute to the total emitted electrons. For the composition-graded $Al_xGa_{1-x}As$ layer, some physical properties, for example, electron mobility (μ), electron diffusion coefficient (D_n), and electron recombination lifetime (τ) are the functions of the Al composition x, which are expressed as follows [26, 27]:

$$\mu = \begin{cases} 8000 - 22,000x + 10,000x^2 \ \left(\text{cm}^2\ \text{V}^{-1}\ \text{S}^{-1}\right), & 0 < x < 0.45 \\ -255 + 1160x - 720x^2 \ \left(\text{cm}^2\ \text{V}^{-1}\ \text{S}^{-1}\right), & 0.45 < x < 1 \end{cases} \tag{5}$$

$$D_n = \begin{cases} 200 - 550x + 250x^2 \ \left(\text{cm}^2\ \text{s}^{-1}\right), & 0 < x < 0.45 \\ -6.4 + 29x - 18x^2 \ \left(\text{cm}^2\ \text{s}^{-1}\right), & 0.45 < x < 1 \end{cases} \tag{6}$$

$$\tau = 29.142 + \frac{4.444 - 29.142}{1 + e^{\frac{x-0.3443}{0.00468}}}, \qquad 0 < \ x \ < 1 \tag{7}$$

Because of the aforesaid variable physical properties regarding to Al composition, the continuity equation of electron transport in the $Al_xGa_{1-x}As$ window layer is quite complex. For simplicity, the $Al_xGa_{1-x}As$ layer is treated to be of a series of sublayers with different Al compositions. As shown in **Figure 2**, the $Al_xGa_{1-x}As$ window layer can be considered to be of n sublayers, wherein T_{wn} denotes the thickness of nth sublayer, and T_{dn} denotes the coordinate point along the y-axis. In this case, the transport of photoelectrons in the $Al_xGa_{1-x}As$ window layer follows the one-dimensional continuity equation through diffusion and drift under the built-in electric field, which is as follows:

$$D_{ni}\frac{d^2n_i(y)}{dy^2} + \mu_i|E_1|\frac{dn_i(y)}{dy} - \frac{n_i(y)}{\tau_i} + g_i(y) = 0, \quad i = 1, 2, 3, \ldots\ldots, n \tag{8}$$

where $g_i(y)$ represents the photoelectron generation function in each $Al_xGa_{1-x}As$ sublayer and is expressed as [28, 29]:

$$g_i(y) = \begin{cases} (1-R_{hv})I_0\alpha_{hv_i}\left[\prod_{m=i+1}^{n}\exp\left(-\alpha_{hv_m}T_{w_m}\right)\right]\exp\left[-\alpha_{hv_i}\left(T_{d_i}-y\right)\right], & i=1,2,\ldots,n-1 \\ (1-R_{hv})I_0\alpha_{hv_i}\exp\left[-\alpha_{hv_i}\left(T_e+\sum_{i=1}^{n}T_{w_i}-y\right)\right], & i=n \end{cases} \tag{9}$$

In Eqs. (8) and (9), i represents the $Al_xGa_{1-x}As$ sublayer along the y axis direction, $n_i(y)$ and α_{hvi} denote the excess electron concentration and the absorption coefficient in each part of $Al_xGa_{1-x}As$ sublayer, I_0 is the incident light intensity, R_{hv} is the reflectivity at the light incident surface, and T_e is the active layer thickness. Besides, the three physical properties, that is, D_{ni}, μ_i, and τ_i in each sublayer are expressed by aforesaid Eqs. (5)–(7).

The excess electron concentration in the former sublayer should contribute to the latter sublayer, accordingly, the boundary conditions adequate for each sublayer are expressed as [28, 29]:

$$\begin{cases} \left[D_{ni}\dfrac{dn_i(y)}{dy}+\mu_i|E_1|n_i(y)\right]\Big|_{y=T_{d_i}} = -S_{v_{i+1}}n_i(y)\Big|_{y=T_{d_i}} + S_{v_{i+1}}n_{i+1}(y)\Big|_{y=T_{d_i}} \\ \left[D_{ni}\dfrac{dn_i(y)}{dy}+\mu_i|E_1|n_i(y)\right]\Big|_{y=T_e} = S_{vi}n_i(y)\big|_{y=T_e} \end{cases}, \quad i=1 \tag{10}$$

$$\begin{cases} \left[D_{ni}\dfrac{dn_i(y)}{dy}+\mu_i|E_1|n_i(y)\right]\Big|_{y=T_{d_i}} = -S_{v_{i+1}}n_i(y)\Big|_{y=T_{d_i}} + S_{v_{i+1}}n_{i+1}(y)\Big|_{y=T_{d_i}} \\ \left[D_{ni}\dfrac{dn_i(y)}{dy}+\mu_i|E_1|n_i(y)\right]\Big|_{y=T_{d_{i-1}}} = S_{vi}n_i(y)\Big|_{y=T_{d_{i-1}}} \end{cases}, \quad i=2,\ldots,n-1 \tag{11}$$

$$\begin{cases} n_i(y)\Big|_{y=T_{d_i}} = 0 \\ \left[D_{ni}\dfrac{dn_i(y)}{dy}+\mu_i|E_1|n_i(y)\right]\Big|_{y=T_{d_{i-1}}} = S_{vi}n_i(y)\Big|_{y=T_{d_{i-1}}} \end{cases}, \quad i=n \tag{12}$$

where S_{vi} is the electron recombination velocity at each interface. By recursively solving the above continuity equations, the excess electron concentration $n_1(T_e)$ reaching the $Al_xGa_{1-x}As$/GaAs interface can be calculated.

As for the GaAs active layer, the excess electrons consist of electrons contributed from the $Al_xGa_{1-x}As$ window layer and electrons generated in the GaAs active layer. Under the second-stage built-in electric field, the photoelectron transport process in GaAs active layer follows the one-dimensional continuity equation as described by

$$D_{n0}\frac{d^2n_0(y)}{dy^2}+\mu_0|E_0|\frac{dn_0(y)}{dy}-\frac{n_0(y)}{\tau_0}+(1-R_{hv})I_0\alpha_{hv0}\left[\prod_{m=1}^{n}\exp\left(-\alpha_{hv_m}T_{w_m}\right)\right]\exp\left[-\alpha_{hv0}(T_e-y)\right]=0, \quad y\in[0,T_e] \tag{13}$$

where $n_0(y)$ is the excess electron concentration in the GaAs active layer, α_{hv0} is the absorption coefficient of the GaAs active layer, and D_{n0}, μ_0, and τ_0 denote the electron diffusion coefficient, the electron mobility, and the electron recombination lifetime in the GaAs active layer, respectively. Considering that the electrons from the $Al_xGa_{1-x}As$ window layer can contribute to the GaAs active layer, the boundary conditions adequate for Eq. (13) are given by [28]:

$$\left\{ \left[D_{n0}\frac{dn_0(y)}{dy} + \mu_0|E_0|n_0(y) \right] \right|_{y=T_e} = -S_{v1}n_0(y)\big|_{y=T_e} + S_{v1}n_1(y)\big|_{y=T_e} n_0(y)\big|_{y=0} = 0 \tag{14}$$

By solving Eq. (13) via the boundary conditions Eq. (14) and the electron concentration $n_1(T_e)$ from the $Al_xGa_{1-x}As$ window layer, the concentration of electrons $n_0(y)$ in the active layer can be figured out. Finally, the quantum efficiency $Y(hv)$, defined as the emitted electron number per incident photon, for the complex $Al_xGa_{1-x}As$/GaAs photocathode is calculated as follows:

$$Y(hv) = PD_{n0}\frac{dn_0(y)}{dy}\Big|_{y=0}/I_0 \tag{15}$$

where P is the surface electron escape probability. If $E_0 = 0$, the quantum efficiency model of $Al_xGa_{1-x}As$/GaAs photocathode with the graded-composition (g-composition) and uniform-doping (u-doping) structure can be obtained. In the same way, when $E_1 = 0$ and $E_0 = 0$, we can also deduce the quantum efficiency model of common t-mode AlGaAs/GaAs photocathodes with the uniform-composition (u-composition) and u-doping structure. In a word, the aforementioned derivation method of photoemission model is applicable to those t-mode photocathodes with a common or complex structure.

Meanwhile, it is noted that the quantum efficiency has a close relation with the reflectivity $R(hv)$ of photocathode, as shown in Eqs. (9) and (13), thus the optical properties of t-mode graded bandgap $Al_xGa_{1-x}As$/GaAs photocathodes need to be investigated. In fact, the usual t-mode photocathode can be treated as a multilayer module, which comprises the glass faceplate, the antireflection layer, the window layer, and the GaAs active layer. The typical structure of t-mode $Al_xGa_{1-x}As$/GaAs photocathodes is shown in **Figure 3**. The glass substrate with a thickness of several millimeters is much thicker than other thin layers in the order of nanometers or micrometers, so the glass is treated as the incident medium rather than the thin film. The reflectivity of incident light permeating the glass substrate is greatly declined by the silicon nitride (Si_3N_4) antireflection film, and then the light in the wave range of interest is absorbed by the $Al_xGa_{1-x}As$ window layer and GaAs active layer in succession. The optical

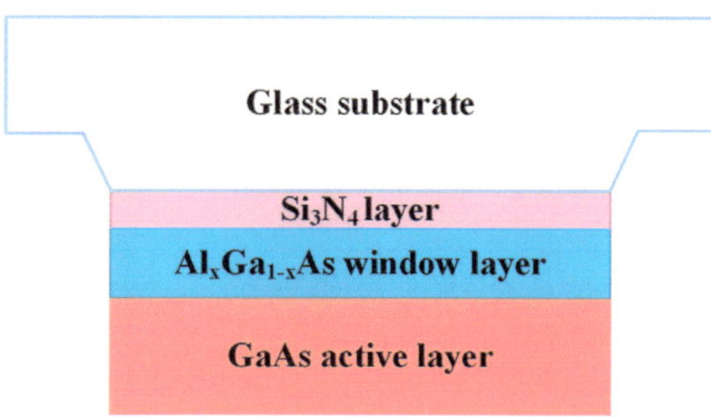

Figure 3. Structural schematic of multilayered t-mode GaAs cathode module, including the glass substrate, the Si_3N_4 antireflection layer, the $Al_xGa_{1-x}As$ window layer, and the GaAs active layer.

properties of multilayer module can be calculated based on the transfer matrix of thin-film optics, and the characteristic matrix of the multilayered cathode module is given by [30]:

$$\begin{bmatrix} B \\ C \end{bmatrix} = \left\{ \prod_{j=1}^{K} \begin{bmatrix} \cos\delta_j & \frac{i}{\eta_j}\sin\delta_j \\ i\eta_j \sin\delta_j & \cos\delta_j \end{bmatrix} \right\} \begin{bmatrix} 1 \\ \eta_{K+1} \end{bmatrix} \tag{16}$$

$$\delta_j = 2\pi\eta_j d_j \cos\theta_j / \lambda \tag{17}$$

$$\eta_j = n_j - ik_j \tag{18}$$

In Eqs. (16)–(18), δ_j and η_j are the optical phase difference and complex refractive index of the jth film layer, η_{K+1} is the optical constant of emergent medium, n_j and k_j constituting the complex refractive index are the refractivity and the extinction coefficient, d_j is the thickness of the jth film layer, and θ_j is the refraction angle of incident light. When light is perpendicularly incident on the surface of the glass substrate, the refraction angle is equal to zero. The reflectivity R_{hv} and transmittivity T_{hv} of the multilayered photocathode module can be calculated by the following expressions [30]

$$R_{hv} = \left(\frac{\eta_g B - C}{\eta_g B + C}\right)\left(\frac{\eta_g B - C}{\eta_g B + C}\right)^* \tag{19}$$

$$T_{hv} = \frac{4\eta_g \eta_{K+1}}{\left(\eta_g B + C\right)\left(\eta_g B + C\right)^*} \tag{20}$$

where η_g denotes the optical constant of the glass substrate. For the Al composition-varied window layer, the optical parameters, for example, the refractivity and extinction coefficient are different in each AlGaAs sublayer [31]. When the $Al_xGa_{1-x}As$ window layer is composed of n sublayers, the t-mode photocathode module can be treated as the thin film system of $n + 2$ layers to calculate the optical properties changing with incident photon wavelength, which are used as the necessary supplement to the quantum efficiency model.

3.2. Quantum efficiency simulation

As to the t-mode AlGaAs/GaAs photocathodes, the optical properties between the g-composition and u-composition structures should be different. For simplified calculation, the composition-graded $Al_xGa_{1-x}As$ window layer is assumed to be of five sublayers with the fixed Al composition in each sublayer. The five Al composition values are assumed to be 0.9, 0.675, 0.45, 0.225, and 0, respectively, distributed from the AlGaAs/Si_3N_4 interface to AlGaAs/GaAs interface. For the u-composition AlGaAs/GaAs photocathode, the Al composition in the AlGaAs window layer is assumed to be 0.7. The optical properties including the reflectivity R_{hv} and transmittivity T_{hv} can be simulated by utilizing Eqs. (16)–(20) by referring to the structure of **Figure 3**. In the simulations, the refractivity and extinction coefficients of $Al_xGa_{1-x}As$ with different Al compositions are referred to [31], the refractivity coefficients of glass and Si_3N_4 are

1.49 and 2.06, respectively, the extinction coefficients of glass and Si_3N_4 are zero due to no absorption, and the thicknesses of Si_3N_4, AlGaAs, and GaAs layer are assumed to be 100, 500 nm and 1.0 μm, respectively. Besides, each sublayer in the g-composition $Al_xGa_{1-x}As$ window layer is supposed to have the equal thickness of 0.1 μm. The simulated optical property curves between the two t-mode AlGaAs/GaAs photocathode modules with different window layer structures are shown in **Figure 4**. It is clear to see that the oscillation number in the entire 400–1100 nm region for the g-composition structure is less than those for the u-composition structure. In other words, compared with the u-composition $Al_{0.7}Ga_{0.3}As$/GaAs photocathode, the g-composition $Al_xGa_{1-x}As$/GaAs photocathode exhibits the much smoother reflectivity curve in the spectrum region of 400–900 nm, which is the concerned photon absorption waveband for the AlGaAs/GaAs material. Besides, in the 900–1100 nm region, the locations of peaks and valleys of the reflectivity curves for the g-composition $Al_xGa_{1-x}As$/GaAs photocathode move toward the shortwave direction, nevertheless, this has little effect on the photoemission performance of GaAs photocathodes since these photons with the wavelength greater than 900 nm are hardly absorbed by the GaAs material.

By using the deduced quantum efficiency models which take into account the reflectivity varying with the wavelength, the quantum efficiency curves of the t-mode $Al_xGa_{1-x}As$/GaAs photocathode with those unique graded bandgap structures are simulated, wherein the window layer is of the g- or u-composition structure, and the active layer is of the e- or u-doping structure, respectively. **Figure 5** exhibits the superiority of the $Al_xGa_{1-x}As$/GaAs photocathode with g-composition window layer and e-doping active layer. In **Figure 5**, some structural parameters such as the Al composition in the u-composition window layer, the Al composition

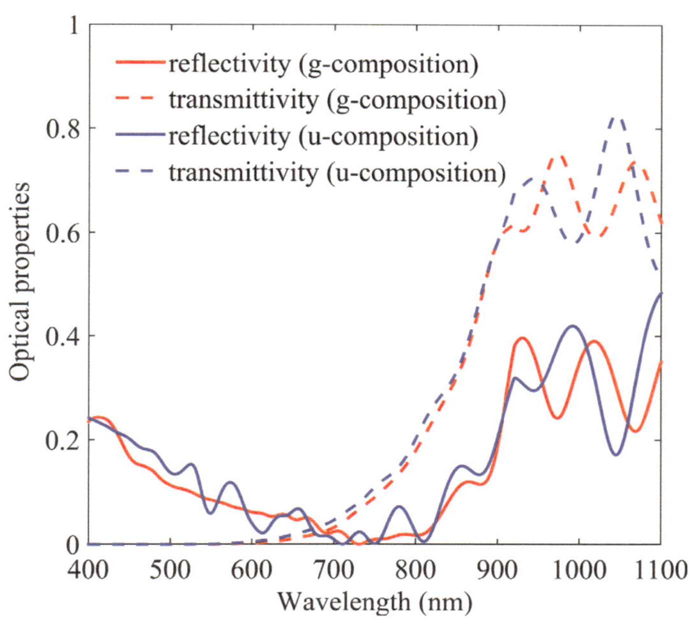

Figure 4. Simulation comparison of optical properties between the two cathode modules with different AlGaAs window layer structures.

distribution in each sublayer of $Al_xGa_{1-x}As$ window layer, and the thicknesses of Si_3N_4, AlGaAs and GaAs layers are identical to those in **Figure 4**. In the GaAs active layer, the doping concentration for e-doping structure is exponentially varied from 1×10^{19} to 1×10^{18} cm^{-3}, and that for the u-doping structure is 1×10^{19} cm^{-3}. In addition, the surface electron escape probability P is assumed to be 0.5. As a result of the reduced lattice mismatch by the seamless $Al_xGa_{1-x}As$/GaAs heterojunction, the interface recombination velocity S_{vi} for g-composition structure cannot exceed 10^4 cm/s, while S_v for the common u-composition structure is usually 10^6 cm/s [5].

It is seen clearly from **Figure 5** that the t-mode g-composition and e-doping photocathode can obtain the highest quantum efficiency in the spectrum region from 400 to 900 nm in contrast to other photocathodes. The quantum efficiency in the shortwave region, that is, blue-green region are enhanced greatly for the two former photocathodes with the g-composition structure. In the g-composition $Al_xGa_{1-x}As$ window layer, the photoelectrons excitated by shortwave light would be promoted toward the GaAs active layer under the g-composition induced electric field. Then, these shortwave photoelectrons are successively boosted toward the emission surface under the built-in electric field formed by the e-doping structure. As shown in **Figure 5**, the e-doping structure for the g-composition $Al_xGa_{1-x}As$/GaAs photocathodes can slightly enhance the quantum efficiency, which is not like the case for the u-composition AlGaAs/GaAs photocathodes. The possible reason is that the g-composition $Al_xGa_{1-x}As$ layer can also absorb some extra longwave photons, which are originally absorbed by the GaAs active layer. In other words, more enough absorption space for longwave photons can be provided by the g-composition structure. While for the u-composition AlGaAs/GaAs

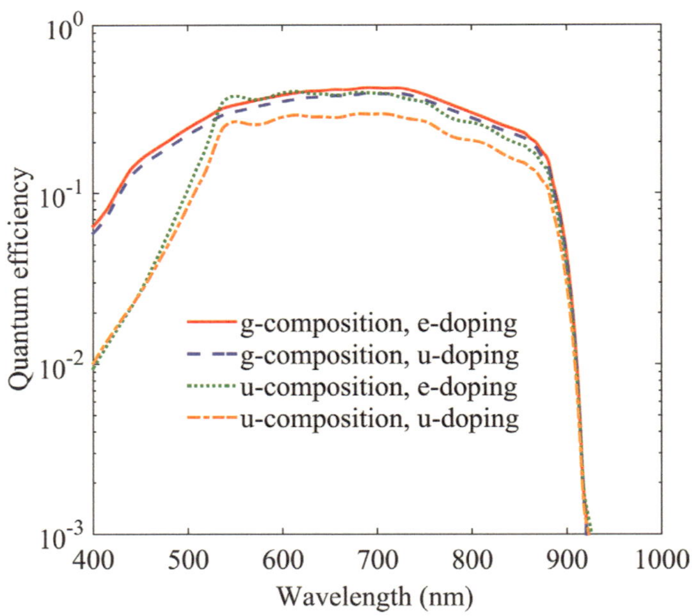

Figure 5. Simulation comparison of quantum efficiency among the t-mode photocathodes with different AlGaAs window layer and GaAs active layer structures.

photocathodes, the case is different. The GaAs active layer just absorbs the longwave photons, and the transport efficiency for these generated photoelectrons can just be improved by the doping-induced electric field.

To guide the structural design of t-mode graded bandgap $Al_xGa_{1-x}As/GaAs$ photocathode, the changes of quantum efficiency with the active layer thickness and the window layer thickness are analyzed, as shown in **Figure 6**. **Figure 6(a)** shows the changes of quantum efficiency curves with the active layer thickness T_e, assuming Al composition distribution, S_{vi} and T_{wi} are the same as those in **Figure 5**. As T_e increases, more space in the bulk for absorption of the longwave photons in the region of 650–900 nm is provided to generate more electrons to increase the quantum efficiency. If the GaAs active layer is thin, photoelectrons generated by shortwave light in the $Al_xGa_{1-x}As$ window layer would easily transport toward the GaAs active layer through diffusion and drift under the two-stage built-in electric field and finally escape into vacuum. In such a case, the quantum efficiency in the shortwave region would remain unchanged. Nevertheless, the thickness of the GaAs active layer must be controlled within a certain range, and the sufficiently thick active layer would decrease the quantum efficiency in the shortwave region, as shown in **Figure 6(a)**. Therefore, the thickness of the active layer should be designed to balance the longwave response and shortwave response. When the $Al_xGa_{1-x}As$ window layer is 500 nm in total thickness, the appropriate thickness is thought to be in the range of 1.0–1.5 μm.

Considering that the built-in electric field in the window layer is inversely proportional to window layer thickness, the effect of the window layer thickness on quantum efficiency in the shortwave region, especially in the blue-green waveband for g-composition photocathodes, is more pronounced than that for the u-composition ones. **Figure 6(b)** shows the quantum efficiency changing with the window layer thickness T_w, assuming T_e = 1.0 μm. As T_w decreases, the quantum efficiency in the waveband region from 400 to 720 nm is greatly enhanced arising from the enhanced g-composition induced electric field. When T_w is thin,

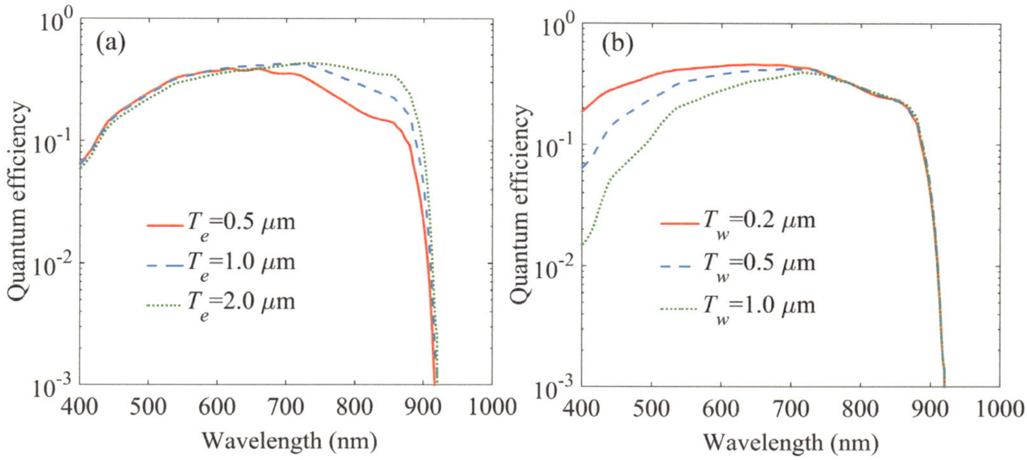

Figure 6. Quantum efficiency simulations with the changes of (a) active layer thickness and (b) window layer thickness for the t-mode graded bandgap $Al_xGa_{1-x}As/GaAs$ photocathodes.

there is not enough space to absorb shortwave photons, and these shortwave photons can be absorbed by GaAs active layer. The quantum efficiency in the shortwave region would get increased as T_w decreases since that the transport capacity of photoelectrons in GaAs layer is better than that in AlGaAs layer. Nevertheless, a passivation layer is necessary to prevent impurities from the substrate into the active layer for the t-mode photocathodes, and thus the $Al_xGa_{1-x}As$ window layer also utilized as the passivation layer should not be extremely thin.

4. Epitaxial growth and quality characterization

4.1. Epitaxial growth of photocathode materials

In modern epitaxial growth techniques, the metalorganic chemical vapor deposition (MOCVD) technique is suitable for growing the complex ultrathin multilayer materials with the composition-graded or doping-graded structures. To confirm the actual effect of the g-composition and e-doping structure on the quantum efficiency of t-mode AlGaAs/GaAs photocathodes, the 2-inch-diameter $Al_xGa_{1-x}As$/GaAs epilayers with two different structures were grown on the low-defect n-type GaAs (100) substrates in the horizontal low-pressure MOCVD reactor from AIXTRON. As shown in **Figure 7(a)**, the multiple epitaxial layers consist of four AlGaAs/GaAs heterostructures, which follow the "inverted structure" technology [32, 33]. In **Figure 7(a)**, the AlGaAs stop layer serves as an etching-resistance layer, and the GaAs cap layer serves as an oxidation-blocking layer. The detailed structures of the two types of cathode materials are shown in **Figure 7(b)** and **(c)**. The difference between the two samples is the structure of window layer, wherein one is of g-composition $Al_xGa_{1-x}As$ layer, and the other is of u-composition $Al_{0.7}Ga_{0.3}As$ layer. Note that, as a result of the current epitaxial limitation, the GaAs active layer exhibits a quasi-exponential doping structure with the p-type dopant concentration varying from 1×10^{19} to 1×10^{18} cm^{-3}.

During the epitaxial growth process of the multiple layers, the group III sources are the trimethylgallium (TMGa) and trimethylaluminum (TMAl), the group V source was the AsH_3, the dopant source was the diethylzinc (DEZn), and the carrier gas was the H_2 gas. Additionally, the growth process was monitored in situ using the LayTech EpiRAS-200 spectrometer. The parameters of the epitaxial growth process are as follows: the growth rate was about 2.5 μm/h, the V/III flux ratio was adjusted at 10–15, the Al composition was controlled by the flow ratio of TMGa to TMAl, and the growth temperature was set as 680°C and 710°C for GaAs and $Al_xGa_{1-x}As$, respectively.

4.2. Quality characterization of photocathode materials

To understand the profile structure of the multilayered photocathode samples, the cross-sectional photographs of the multilayered structure for the two cathode material samples were measured by the scanning electron microscope (SEM) from Hitachi. It is clearly seen from **Figure 8** that differing from the case for u-composition sample, no sharp borderline exists at the interface of the $Al_xGa_{1-x}As$ window layer and GaAs active layer for the g-composition sample. This seamless interface would greatly reduce the interface electron recombination. It is

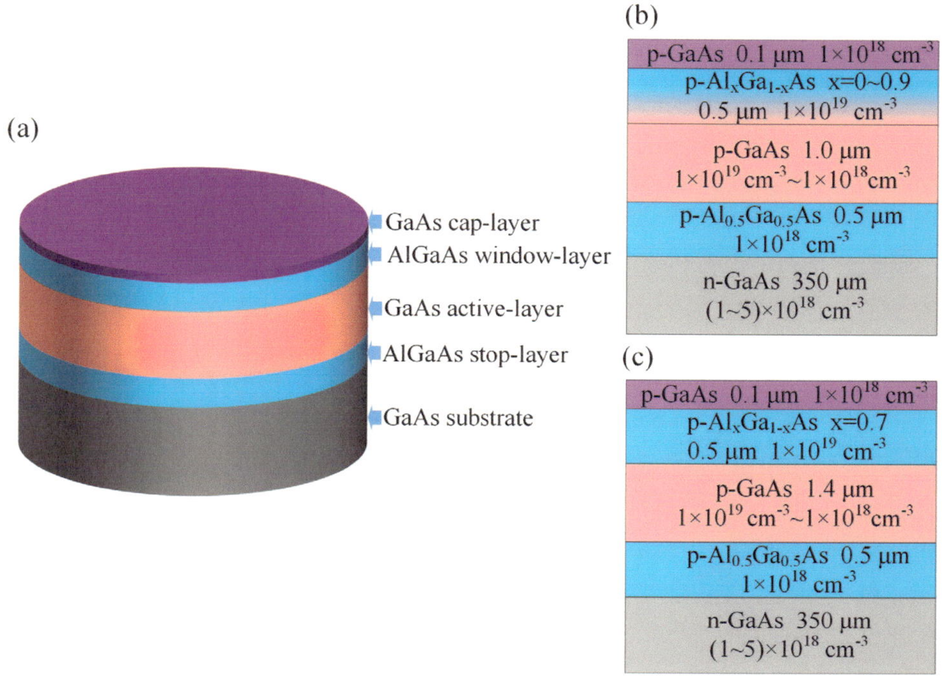

Figure 7. (a) Schematic diagram of the epitaxial t-mode AlGaAs/GaAs photocathode materials following the "inverted structure" technology, the detailed epitaxial structures of (b) g-composition and e-doping cathode sample, and (c) u-composition and e-doping cathode sample.

noted that many cracks in **Figure 8(b)** are caused by the inappropriate cleavage, which cannot reflect the true quality of the epitaxy. From the SEM photographs, it is judged that the vertically multilayered constructions of the epitaxial cathode materials agree well with the structural design.

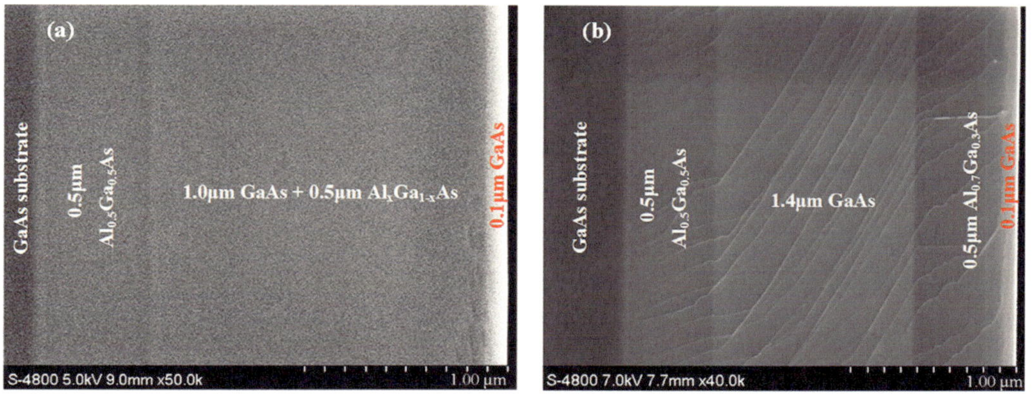

Figure 8. Cross-sectional SEM photographs of the cleaved epitaxial cathode samples with (a) g-composition and e-doping structure and (b) u-composition and e-doping structure.

The depth distribution of carrier concentration in the multilayered p-type AlGaAs/GaAs materials was measured by the electrochemical capacitance-voltage (ECV) system from Bio-Rad. As shown in **Figure 9**, a series of sublayers forming the graded doping structure can be realized by the MOCVD technique. The carrier concentration of no more than 8 × 10^{18} cm^{-3} in the GaAs active layer shows a gradient distribution. For the $Al_xGa_{1-x}As$ window layer in **Figure 9 (a)**, the carrier concentration decreases with the increase in Al composition, which exactly reflects the composition-graded structure.

To investigate the crystalline quality of the epitaxial cathode materials, the X-ray diffraction (XRD) curves were measured by the X'Pert Pro MRD system. As shown in **Figure 10**, the

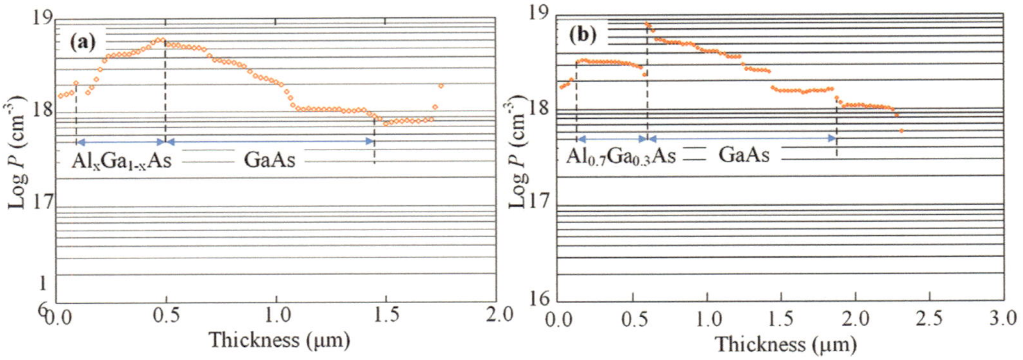

Figure 9. Depth distribution of carrier concentration in the cleaved epitaxial cathode samples with (a) g-composition and e-doping structure and (b) u-composition and e-doping structure.

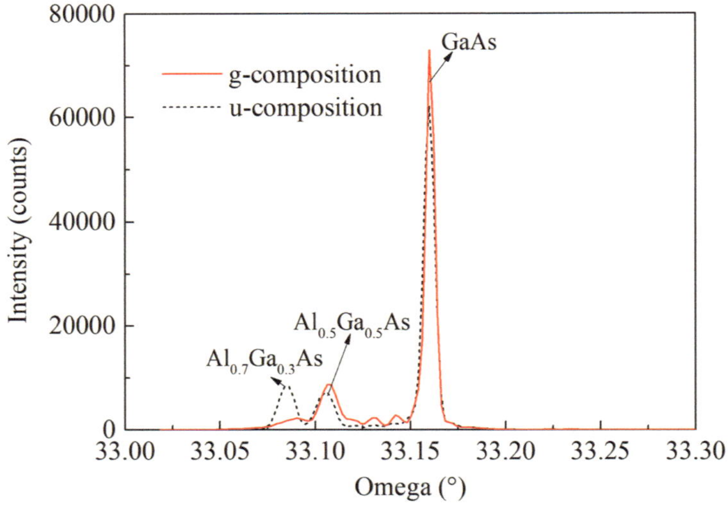

Figure 10. XRD curves of the cleaved epitaxial cathode samples with two different structures.

rightmost peak represents the GaAs material, which is the superposition of the diffraction peaks of the GaAs cap layer, active layer, and substrate. The only one diffraction peak indicates that the crystalline perfection of the GaAs epilayers is consistent with the GaAs substrate. The left two diffraction peaks for the u-composition sample represent the AlGaAs window layer and stop layer, respectively. In the g-composition sample, there is no diffraction peak denoting the window layer, and a series of diffraction peaks exist nearby the peak of the GaAs layer, which are caused by the g-composition $Al_xGa_{1-x}As$ epilayer. The slightly narrower full width at half maximum of the GaAs diffraction peak indicates that the GaAs active layer in the g-composition sample has a better crystalline quality.

5. Device fabrication and spectral response

5.1. Transmission-mode cathode module fabrication

Following the recipe of fabricating glass-sealed t-mode AlGaAs/GaAs photocathodes [32, 33], the epitaxial cathode materials cutted from the 2-inch-diameter epitaxial wafer were fabricated into the multilayered t-mode cathode module. The schematic process flow for fabricating t-mode AlGaAs/GaAs photocathode modules is shown in **Figure 11**. First, the GaAs cap layer was removed by chemical etching to expose the AlGaAs window layer, and by plasma enhanced chemical vapor deposition (PECVD), a thin antireflective layer of 100 nm-thick Si_3N_4 was deposited on the exposed window layer surface. Then, the 7056 glass, serving as the incident window and support layer, was bonded on the Si_3N_4 antireflection layer by thermocompression. Following that, through selective etching process, the GaAs substrate and AlGaAs stop layer were etched away to expose the GaAs active layer to prepare the NEA surface [32]. Finally, the Cr-Ni ring electrode applied to bias on the cathode was prepared by the physical vapor deposition (PVD), such as magnetron sputtering method. After these processing steps, the multilayered cathode module with a glass/Si_3N_4/AlGaAs/GaAs structure was finished. In addition, to eliminate etching-induced damage at the active layer surface, the polishing treatment was implemented, which slightly decreased the thickness of the GaAs active layer.

The optical property curves of the t-mode cathode modules with two different structures were measured by utilizing the Shimadzu UV-3600 spectrophotometer, which possesses three detectors working from ultraviolet to NIR waveband. The optical properties were measured based on the double optical path method, and light was incident on the surface of glass faceplate in a normal direction. **Figure 12** shows the experimental reflectivity and transmissivity curves of the two different multilayered module samples. It is found that, just as the simulated results in **Figure 4**, the reflectivity curve in the region of 400–800 nm for g-composition structure is relatively smoother than that for u-composition structure. In other words, the smooth reflectivity curve verifies the composition-graded structure in the $Al_xGa_{1-x}As$ window layer from another aspect. Thereby, the characterization results regarding the cross-sectional photographs, carrier concentration distributions, X-ray diffraction peaks, and optical properties all reflect the special design structure.

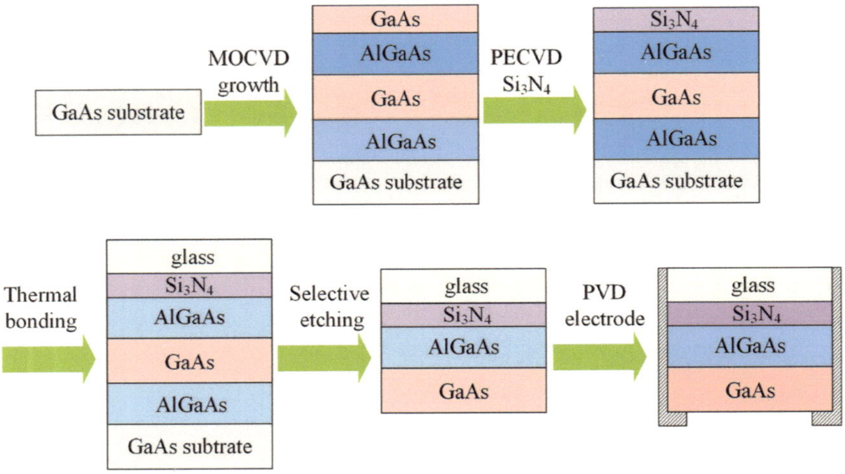

Figure 11. Schematic of the process flow for fabricating t-mode AlGaAs/GaAs photocathode modules following the "inverted structure" technology.

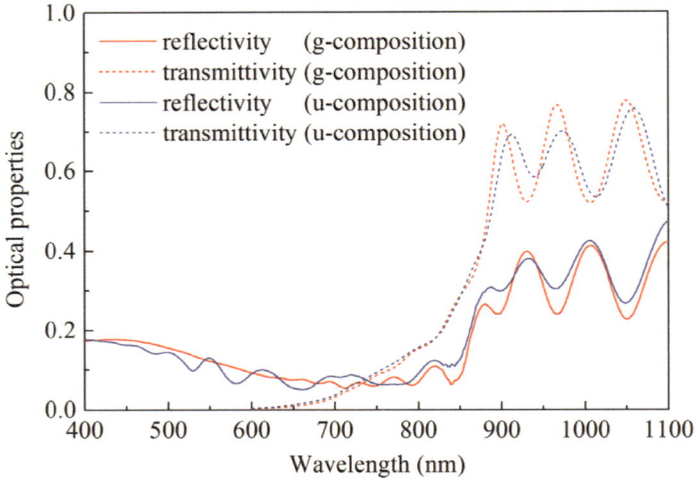

Figure 12. Experimental optical property curves of the t-mode AlGaAs/GaAs cathode modules with two different structures.

5.2. Activation of photocathode surface

Prior to activation, the 18-mm-diameter cathode modules experienced the chemical cleaning and vacuum annealing to obtain an atomic level clean surface. The heat treatment with a suitable temperature under ultrahigh vacuum (UHV) condition is particularly important for the activation, and the quadrupole mass spectrometer (QMS) was adopted to monitor the change of residual gas components during the programmed temperature rose and fell. **Figure 13** shows the changes of mainly concerned residual gas components for the two t-mode cathode module samples. Through detecting the gas presence of the QMS traces at *m*/*e* = 18 (H_2O), 75 (As), 91 (AsO), 150 (As_2), and 156 (Ga_2O), it can be judged that whether the oxides on the GaAs surface, such as As_2O_3 and Ga_2O_3, are cleared away with the increased temperature or not [34]. It can be inferred from **Figure 13** that both the cathode module samples obtained an oxide-free clean surface after the heat treatment procedure in terms of these obvious QMS trace peaks.

After the sample cooled to room temperature, the Cs—O activation to form the NEA state at the cathode surface was performed in the UHV chamber with a base pressure of 10^{-9} Pa. The Cs and O sources used in the activation are solid dispensers easily controlled by direct current, and the flux is proportional to the operating current [35]. During the activation, the Cs source was on all the time, and the O source was switched on and off [35]. The operating current of Cs and O dispensers was regulated by program control current supply, and the photocurrent induced by a white light source was monitored in real time by the computer-controlled test system [35]. The initial Cs supply caused the gradual increase of the photocurrent. With the continuous Cs flux, when the photocurrent dropped to 80% of its peak, the O source was open. In subsequent alternate activation cycles, the O source was closed when the photocurrent reached its peak and was open again when the photocurrent dropped to 80% of the peak. The operating current ratio of Cs source to O source for both samples was regulated as the same 1.65/1.8. Until the photocurrent peak no longer increased, the O source and Cs source were closed successively, and the activation process was finished. To further improve the photoemission performance, the second heat treatment with a lower temperature was employed to the samples [36]. After that, the samples were activated again using the same co-deposition activation. As seen from **Figure 14**, the second activation can dramatically enhance the final cathode performance. Meanwhile, the final photocurrent peaks of the two samples are approximately the same.

5.3. Tube package and spectral response test

After the two-step Cs—O activation process, the cathode module in the UHV activation chamber was transferred to the UHV seal vacuum chamber and indium sealed into an image

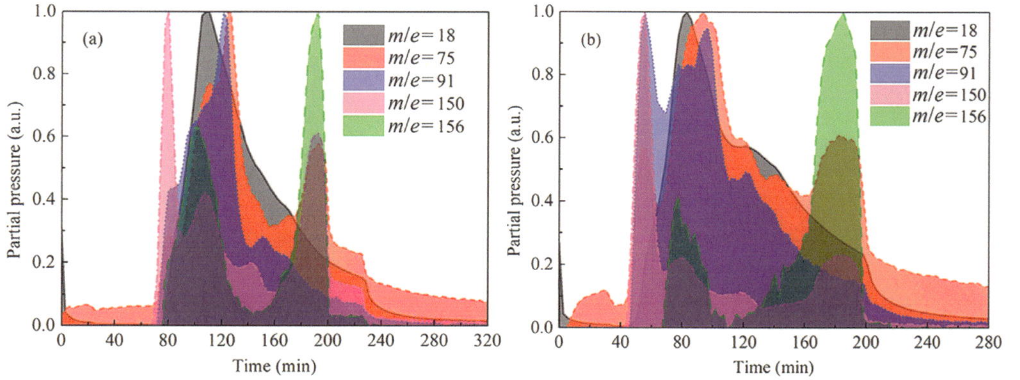

Figure 13. Residual gas changes during high-temperature thermal cleaning process for (a) g-composition and (b) u-composition AlGaAs/GaAs cathode modules.

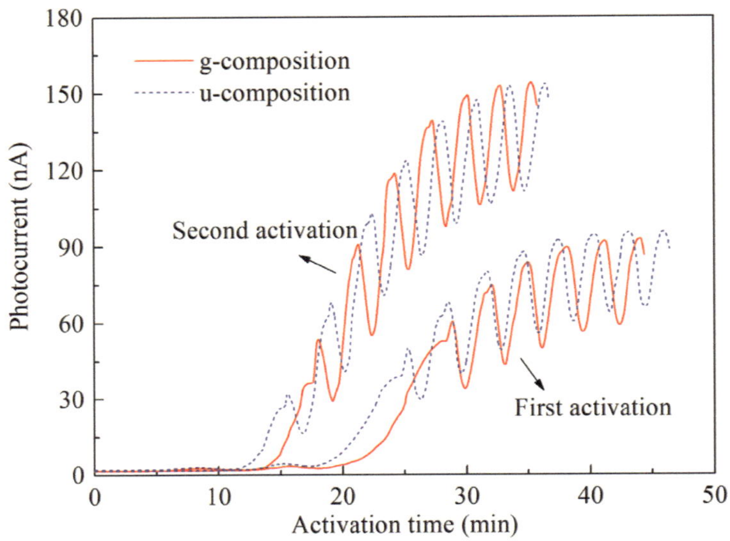

Figure 14. Photocurrent changes during Cs—O activation for the two t-mode AlGaAs/GaAs cathode modules.

intensifier tube, wherein the t-mode AlGaAs/GaAs cathode module was equipped in association with the filmed microchannel plate (MCP), phosphor screen, output window, ceramics, and Kovar sealing parts [37]. The schematic structure and the photograph of the LLL proximity focused image intensifier are shown in **Figure 15**. As shown in **Figure 15(a)**, the proximity focused image intensifier is capable of enhancing a LLL image from several thousands to tens of thousands of times. The input LLL image is converted into photoelectrons by the AlGaAs/GaAs photocathode, and then the number of photoelectrons is multiplied several thousands of times by the MCP coated with a thin ion barrier film which can prevent ion feedback. Lastly, the multiplied photoelectrons bombard the phosphor screen and are converted into photons.

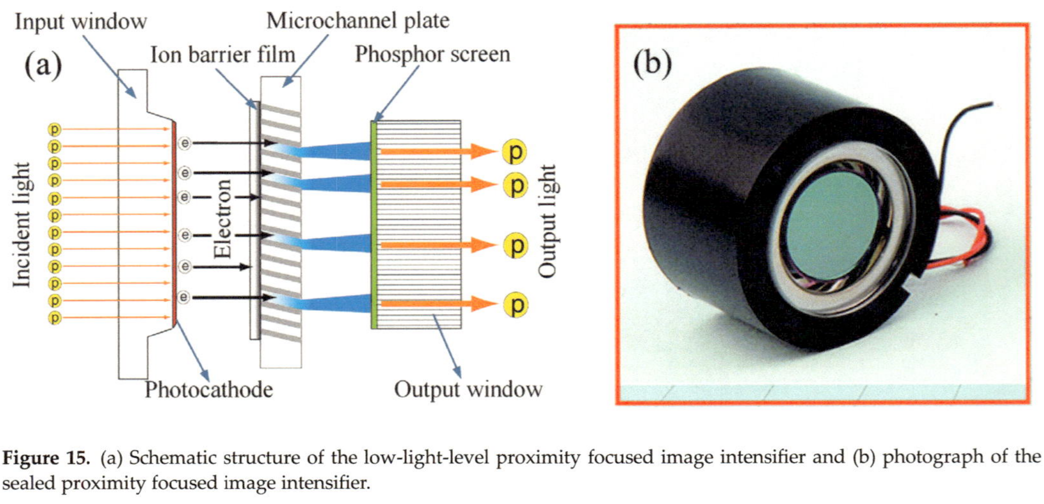

Figure 15. (a) Schematic structure of the low-light-level proximity focused image intensifier and (b) photograph of the sealed proximity focused image intensifier.

Thus, the input LLL image is intensified and appears as the output image on the phosphor screen. In addition to the function of direct eye observation, the LLL image intensifier can be coupled with CCD/CMOS array by the fiber optic taper to realize video output and remote monitoring [38, 39].

The sealed image intensifiers were extracted from the seal vacuum chamber into ambient air, and the spectral response curves were measured by the spectral response testing instrument [22]. Through the spectral response values corresponding to the wavelength, the quantum efficiency values corresponding to the wavelength for the two different cathode samples were obtained [40]. In the spectral region of 600–750 nm, the quantum efficiency exceeds 40%. As shown in **Figure 16**, it is found that in contrast to the u-composition structure, the g-composition structure is especially useful to the enhancement of shortwave quantum efficiency, which conforms to the original intention of our design concept. By fitting the experimental optical property and quantum efficiency data based on the theoretical photoemission model, the internal cathode parameters difficult to be measured directly can be obtained. The thickness values of each layer calculated by fitting the experimental reflectivity and transmittivity curves are listed in **Table 1**. It is seen that the Al composition in the g-composition window layer is not distributed uniformly, and the sublayers with low Al composition are relatively thinner compared to those with high Al composition. For the two samples, the thicknesses of the GaAs active layer are smaller than the design values, which are caused by the polishing treatment after the fabrication of cathode modules.

By means of fitting the experimental quantum efficiency curves, we can obtain some performance parameters, for example, interface recombination velocity S_v and surface escape probability P. It is seen from **Table 1** that the two samples have the same P, which means that the

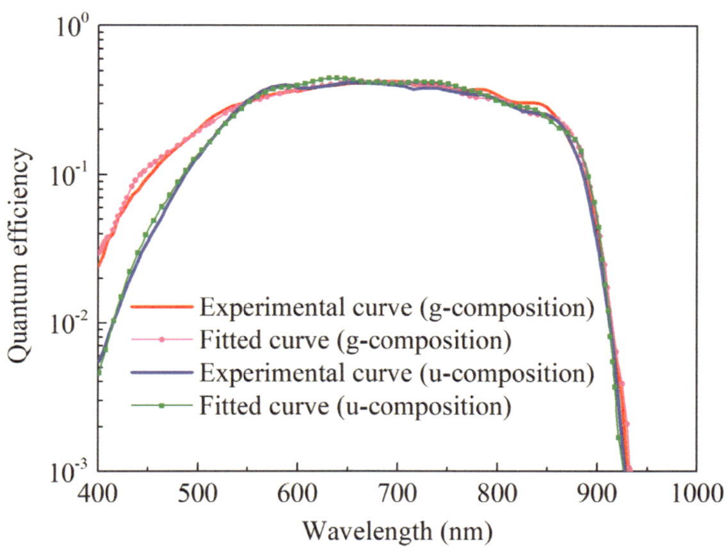

Figure 16. Experimental and fitted quantum efficiency curves for the two different t-mode AlGaAs/GaAs photocathodes samples.

Cathode sample	d_{Si3N4} (nm)	Al composition in each AlGaAs sublayer	d_{AlGaAs} (nm)	d_{GaAs} (nm)	S_v (cm/s)	P
g-Composition	109	0.9	106	857	10^3	0.52
		0.675	117			
		0.45	125			
		0.225	63			
		0	72			
u-Composition	107	0.7	485	1256	10^5	0.52

Table 1. Fitted parameters of the two different t-mode AlGaAs/GaAs photocathode samples.

surface barrier shapes changed by the activation process are the same. Whereas, S_v for the g-composition sample, it is considerably reduced in contrast to that for the u-composition sample. It is known that the S_v is mainly determined by the crystal quality of photocathode itself, such as misfit dislocations and stacking faults at the AlGaAs/GaAs interface. The g-composition structure in the window layer can not only mitigate the interface discontinuity caused by the interface lattice mismatch, but also form an internal electric field to facilitate the transport of shortwave photons excitated electrons toward the emitting surface.

6. Conclusions

In this chapter, we have carried out systematically theoretical and experimental researches on t-mode AlGaAs/GaAs photocathodes, with regard to bandgap structure design, photoemission model derivation, epitaxial growth, surface activation, device fabrication, and performance evaluation. Compared with the common t-mode AlGaAs/GaAs photocathode, the graded bandgap t-mode $Al_xGa_{1-x}As/GaAs$ photocathode with a g-composition and e-doping structure can achieve higher quantum efficiency in the shortwave response region, particularly the blue-green spectral region of interest. In addition, this g-composition structure is helpful to mitigate the interface recombination and enhance the absorption of the longwave light, which leads to the enhanced photoemission capability. This work has reference significance for the design of other graded bandgap III-V group photocathodes.

Acknowledgements

This work was supported by the National Natural Science Foundation of China (grant nos. 61771245 and 61301023) and Science and Technology on Low-Light-Level Night Vision Laboratory Foundation of China (grant no. J20150702). The authors would like to thank Dr. Feng Cheng for her efforts in the theoretical photoemission model, and the staff from Science and Technology on Low-Light-Level Night Vision Laboratory for their assistance in the fabrication of t-mode cathode modules and image intensifiers.

Author details

Yijun Zhang[1]* and Gangcheng Jiao[2]

*Address all correspondence to: zhangyijun423@njust.edu.cn

1 School of Electronic and Optical Engineering, Nanjing University of Science and Technology, Nanjing, China

2 Science and Technology on Low-Light-Level Night Vision Laboratory, Xi'an, China

References

[1] Scheer JJ, Laar JV. GaAs-Cs: A new type of photoemitter. Solid State Communications. 1965;**3**:189-193

[2] Chrzanowski K. Review of night vision technology. Opto-Electronics Review. 2013;**21**:153-181

[3] Liu W, Chen Y, Lu W, Moy A, Poelker M, Stutzman M, et al. Record-level quantum efficiency from a high polarization strained GaAs/GaAsP superlattice photocathode with distributed Bragg reflector. Applied Physics Letters. 2016;**109**:252104

[4] Baum A, Arcuni P, Aebi V, Presley S, Elder M. Prototype negative electron affinity-based multibeam electron gun for lithography and microscopy. Journal of Vacuum Science and Technology B. 1999;**17**:2819-2822

[5] Schwede JW, Sarmiento T, Narasimhan VK, Rosenthal SJ, Riley DC, Schmitt F, et al. Photon-enhanced thermionic emission from heterostructures with low interface recombination. Nature Communications. 2013;**4**:1576

[6] Williams BF, Tietjen JJ. Current status of negative electron affinity devices. Proceedings of the IEEE. 1971;**59**:1489-1497

[7] Karkare S, Boulet L, Cultrera L, Dunham B, Liu X, Schaff W, et al. Ultrabright and ultrafast III–V semiconductor photocathodes. Physical Review Letters. 2014;**112**:097601

[8] Kuwahara M, Takeda Y, Saitoh K, Ujihara T, Asano H, Nakanishi T, et al. Development of spin-polarized transmission electron microscope. Journal of Physics: Conference Series. 2011;**298**:012016

[9] Mitsuno K, Masuzawa T, Hatanaka Y, Neo Y, Mimura H. Activation process of GaAs NEA photocathode and its spectral sensitivity. In: 3rd International Conference on Nanotechnologies and Biomedical Engineering. Singapore: Springer; 2016. pp. 163-166

[10] Martinelli RU, Fisher DG. The application of semiconductors with negative electron affinity surfaces to electron emission devices. Proceedings of the IEEE. 1974;**62**:1339-1360

[11] Antypas GA, James LW, Uebbing JJ. Operation of III-V semiconductor photocathodes in the semitransparent mode. Journal of Applied Physics. 1970;**41**:2888-2894

[12] Estrera JP, Ostromek T, Bacarella A, Isbell W, Iosue MJ, Saldana M, et al. Advanced image intensifier night vision system technologies: Status and summary 2002. Proceedings of SPIE. 2003;**4796**:49-59

[13] Jin X, Yamamoto N, Nakagawa Y, Mano A, Kato T, Tanioku M, et al. Super-high brightness and high-spin-polarization photocathode. Applied Physics Express. 2008;**1**:045002

[14] Pastuszka S, Hoppe M, Kratzmann D, Schwalm D, Wolf A, Jaroshevich AS, et al. Preparation and performance of transmission-mode GaAs photocathodes as sources for cold dc electron beams. Journal of Applied Physics. 2000;**88**:6788-6800

[15] Spicer WE, Herreragomez A. Modern theory and applications of photocathodes. Proceedings of SPIE. 1993;**2022**:18-35

[16] Fisher DG, Enstrom RE, Escher JS, Gossenberger HF, Appert JR. Photoemission characteristics of transmission-mode negative electron affinity GaAs and (ln,Ga)As vapor-grown structures. IEEE Transactions on Electron Devices. 1974;**21**:641-649

[17] Sinor TW, Estrera JP, Phillips DL, Rector MK. Extended blue GaAs image intensifiers. Proceedings of SPIE. 1995;**2551**:130-134

[18] Zhang S, Benson SV, H-Garcia C. Observation and measurement of temperature rise and distribution on GaAs photo-cathode wafer with a 532 nm drive laser and a thermal imaging camera. Nuclear Instruments and Methods in Physics Research. 2011;**631**:22-25

[19] Spicer WE. Negative affinity 3-5 photocathodes: Their physics and technology. Applied Physics. 1977;**12**:115-130

[20] Costello K, Aebi V, Davis G, Rue RL, Weiss R. Transferred electron photocathode with greater than 20% quantum efficiency beyond 1 micron. Proceedings of SPIE. 1995;**2550**: 177-188

[21] Jones LB, Rozhkov SA, Bakin VV, Kosolobov SN, Militsyn BL, Scheibler HE, et al. Cooled transmission-mode NEA-photocathode with a band-graded active layer for high brightness electron source. AIP Conference Proceedings. 2008;**1149**:1057-1061

[22] Zhang YJ, Niu J, Zhao J, Chang BK, Shi F, Cheng HC. Influence of exponential-doping structure on photoemission capability of transmission-mode GaAs photocathodes. Journal of Applied Physics. 2010;**108**:093108

[23] Zhang YJ, Chang BK, Niu J, Zhao J, Zou JJ, Shi F, et al. High-efficiency graded band-gap $Al_xGa_{1-x}As/GaAs$ photocathodes grown by metalorganic chemical vapor deposition. Applied Physics Letters. 2011;**99**:101104

[24] Feng C, Zhang YJ, Qian Y, Chang BK, Shi F, Jiao GC, et al. Photoemission from advanced heterostructured $Al_xGa_{1-x}As/GaAs$ photocathodes under multilevel built-in electric field. Optics Express. 2015;**23**:19478-19488

[25] Yang Y, Yang W, Sun C. Heterostructured cathode with graded bandgap window-layer for photon-enhanced thermionic emission solar energy converters. Solar Energy Materials & Solar Cells. 2015;**132**:410-417

[26] Levinshtein M, Shur MS, Rumyanstev S, editors. Handbook Series on Semiconductor Parameters. Vol. 2. London: World Scientific; 1999

[27] Zarem HA, Lebens JA, Nordstrom KB, Sercel PC, Sanders S, Eng LE, et al. Effect of Al mole fraction on carrier diffusion lengths and lifetimes in $Al_xGa_{1-x}As$. Applied Physics Letters. 1989;**55**:2622-2624

[28] Feng C, Zhang YJ, Qian YS, Xu Y, Liu XX, Jiao GC. Quantum efficiency of transmission-mode $Al_xGa_{1-x}As$/GaAs photocathodes with graded-composition and exponential-doping structure. Optics Communications. 2016;**369**:50-55

[29] Feng C, Zhang YJ, Qian YS, Wang ZH, Liu J, Chang BK, et al. High-efficiency $Al_xGa_{1-x}As$/GaAs cathode for photon-enhanced thermionic emission solar energy converters. Optics Communications. 2018;**413**:1-7

[30] Zhao J, Xiong YJ, Chang BK, Zhang YJ, Zhang JJ. Research on optical properties of transmission-mode GaAs photocathode module. Proceedings of SPIE. 2011;**8194**:81940J

[31] Aspnes DE, Kelso SM, Logan RA, Bhat R. Optical properties of $Al_xGa_{1-x}As$. Journal of Applied Physics. 1986;**60**:754-767

[32] Antypas GA, Edgecumbe J. Glass-sealed GaAs-AlGaAs transmission photocathode. Applied Physics Letters. 1975;**26**:371-372

[33] André JP, Guittard P, Hallais J, Piaget C. GaAs photocathodes for low light level imaging. Journal of Crystal Growth. 1981;**55**:235-245

[34] Yamada M, Ide Y. Anomalous behaviors observed in the isothermal desorption of GaAs surface oxides. Surface Science. 1995;**339**:L914-L918

[35] Zhang YJ, Qian YS, Feng C, Shi F, Cheng HC, Zou JJ, et al. Improved activation technique for preparing high-efficiency GaAs photocathodes. Optical Materials Express. 2017;**7**:3456

[36] Rodway DC, Allenson MB. In situ surface study of the activating layer on GaAs (Cs, O) photocathodes. Journal of Physics D: Applied Physics. 1986;**19**:1353-1371

[37] Thomas N. System performance advances of 18-mm and 16-mm subminiature image intensifier sensors. Proceedings of SPIE. 2000;**4128**:54-64

[38] Nützel G. Single-photon imaging using electron multiplication in vacuum. In: Seitz P, Theuwissen AJP, editors. Single-Photon Imaging. Berlin: Springer; 2011. pp. 73-102

[39] Vallerga JV, Siegmund O, Dalcomo J, Jelinsky PN. High-resolution (<10 μm) photon-counting intensified CCD. Proceedings of SPIE. 1997;**3019**:156-167

[40] Zhang YJ, Zou JJ, Niu J, Chang BK, Xiong YJ. Variation of spectral response for exponential-doped transmission-mode GaAs photocathodes in the preparation process. Applied Optics. 2010;**49**:3935-3940

Chapter 5

a-Si:H p-i-n Photodiode as a Biosensor

Vera Gradišnik and Darko Gumbarević

Additional information is available at the end of the chapter

http://dx.doi.org/10.5772/intechopen.80503

Abstract

The p-i-n a-Si:H photodiode is a promising device as a transducer in biosensors. The native and light-induced localized state density and energy distribution in the energy gap of a-Si:H have a large effect on the photoconductivity of thin-film photodiodes. Depending on their nature, they play a crucial role in trapping and recombination processes and consequently influence the photodiode capacitance. The optical bias dependence of modulated photocurrent, OBMPC, method using the blue LED light is applied to clarify the nature and energy distribution of the energy gap density of states and their influence on the photodiode capacitance, from photodiodes transient response. It is observed that the deep defect states of the i-layer contribute to the capacitance at various bias voltages. Also, the capacitance achieves the upper limit around the built-in potential. Based on this method and obtained results, the a-Si:H p-i-n photodiode is used as a biosensor transducer in the detection of mammalian cell chemiluminescence.

Keywords: a-Si:H p-i-n photodiode, biosensor, blue light, capacitance, defects, density of states, LED, transient response

1. Introduction

The recent advances, miniaturization and integration, in nanotechnology and CMOS technology afforded by photolithographic patterning, have had a transformative impact on the field of single-cell biology and diseases that depend on small collections of cells in their initial stages such as cancer. The microfluidic Lab-on-a-chip technology, still under development, meets point-of-care (POC) requirements for biomolecular analyses. The biosensors consisting of amorphous silicon (a-Si:H) p-i-n photodiode as integrated luminescence sensor in lab-on-a-chip devices, coupled with a microLED, have progressed rapidly over the last two decades and are still under development [1, 2]. The a-Si:H p-i-n photodiode is widely used as a transducer in biosensors for biochemical analysis, where are applied pico- to nano-liters (microliters) of volumes of fluids in channels of tens to hundreds of micrometers. The photodiode array must have very high detection precision and allow conducting parallel experiments for the detection of biomolecules. Biosensor response must satisfy the main performance criteria: selectivity, sensitivity, linearity, and response time. At the same time, the photonic method for measuring the oxygen consumption rate (OCR) of a single cell must be developed.

The sensitivity of thin-film a-Si:H p-i-n photodiodes, integrated with microfluidics, allows low-level luminescence signal detection from the volume of a microfluidic channel. The thin-film hydrogenated amorphous silicon (a-Si:H) technology [3] allows the custom fit of amorphous silicon photodiode arrays to the geometry of the flow microfluidic channel. The low-temperature (below 200°C) technology plasma-enhanced chemical-vapor deposition (PECVD) [2] or hot wire chemical-vapor deposition (HWCVD) [4] allows deposition of amorphous layers on the glass and polymer substrates, respectively, and on top of crystalline silicon integrated circuits without any damage to the circuits below [5]. At appropriate RF power, gas flows, chamber pressure, and substrate temperature in PECVD, hydrogen atoms are introduced into the thin film to passivate the silicon dangling bonds (DBs) and remove a part of (metastable) defect states from the forbidden band gap. In pure amorphous silicon, unsaturated dangling bonds (DBs) give rise to electronic states inside the band gap. The hydrogen atoms restore the energy gap and semiconductor properties. Due to the disordered structure alloying virtually all optical transitions, the absorption coefficient of a-Si:H is higher than that of c-Si (500–650 nm) [6]. Besides, a much lower dark current of a-Si:H than c-Si at room temperature enables its use as a photodiode material for low-noise detection. The photodiodes, as part of active area in active pixel sensors (APSs) [7–10] and other devices based on amorphous silicon, recently entered the field of microelectronics. The main part of applications was directed toward steady-state illumination of slowly varying light signals. The transient photocurrent was used for the material properties characterization and color detection [11–13].

The amorphous silicon photodiode can operate in integrated and in a reverse-biased photodiode mode. In the latter, they have a high response speed and the photocurrent is only controlled by the light intensity. In amorphous silicon, the transport of free carriers involves trapping, detrapping through a large density of midgap states (DOS), and motion through transport in the extended states—localized band tail state [14]. These native and light-induced densities of state and their energy distribution in the energy gap of a-Si:H have a large effect on the photoconductivity of thin-film photodiodes. Depending on their nature, they play a crucial role in trapping and recombination processes and consequently influence the photodiode capacitance and relaxation time. Furthermore, they lead to a high RC constant of a thin-film a-Si:H photodiode.

The disordered structure of hydrogenated amorphous silicon (a-Si:H) leads to localized states as band tails that extend inside the energy gap. The coordination defects associated with dangling bonds are sources of defect states located around the midgap. The tail states are shallow states, and the dangling bonds, the deep states. Both of them influence the recombination processes, capture and reemission of carriers in semiconductor. The emission of free charge carriers from deep states at the p-i and i-n interfaces influences the dark current, space charge inside i-layer, and capacitance. The concentration of midgap states and their spatial distribution in the i-layer and at p^+-i and n^+-i interfaces can be extracted from transient dark current and steady-state thermal generation current, as described by Murthy and Dutta, and by Mahmood and Kabir [15, 16]. Models of transport and recombination through localized states have also been well described by Fuhs [14] and Dhariwal et al. [17–19]. Several techniques based on steady-state and transient photocurrent techniques have been developed to determine the nature and role of gap density of states (DOS) in the trapping-detrapping, recombination processes of mobile carriers and gap-state parameters [20]. To estimate the DOS in the lower part of band gap between the Fermi level and valence band edge, methods such as constant photocurrent method (CPM) [21], Fourier transform photocurrent spectroscopy (FTPS) [22, 23], and dual beam photoconductivity (DBP) [24] were used in the past. On the other hand, the multiexponential trapping rate and modulated photocurrent (MPC) technique [25–27] allow determining parameters of localized states throughout the entire energy gap by employing frequency and temperature scans.

The aim of our research is the mammalian cells chemiluminescence detection, which is based on the phenomenon that under illumination of two-beam, low intensity probe beam and simultaneously a higher intensity bias beam, reverse-biased a-Si:H p-i-n photodiode photocurrent exceed expected primary photocurrent [26].

The transient response of a-Si:H p-i-n photodiode to blue LED light pulse superimposed to the blue LED light optical bias (optical bias dependence of modulated photocurrent method—OBMPC [11, 27]) at various reverse bias voltages and one frequency is applied to clarify the nature and energy distribution of energy gap density of state and their influence on the photodiode capacitance [28]. It is observed that the deep defect states of the i-layer contribute to the capacitance at various bias voltages. Also, the capacitance achieves the upper limit around the built-in potential.

Based on this method and obtained results, we describe our experiment, where the a-Si:H p-i-n photodiode is used as a biosensor transducer in the detection of mammalian cell's chemiluminescence.

2. a-Si:H p-i-n photodiode

2.1. Device structure and characterization

The fundamental structure of a photodiode in amorphous silicon is p-i-n or n-i-p. The a-Si:H p-i-n structure (**Figure 1**) investigated in this work (Sunčane ćelije d.o.o. Split, Croatia) was deposited on a transparent conductive oxide (TCO)-coated glass from undiluted SiH_4 by plasma-enhanced CVD at 13.56 MHz. The different layers of the p-i-n structure have the parameters of standard solar cell production. The thicknesses of the n-type, i-type, and p-type layers were 5, 300, and 5 nm, from top to the bottom, respectively. The n-type layer was made by adding phosphine PH_3 and the p-type by adding diborane B_2H_6 into the silane SiH_4 source gas during growth. The back contact was aluminum deposited by evaporation.

Figure 1. The a-Si:H p-i-n photodiode structure.

The active area of the pixel was 0.81 cm^2. The basic device characterization and experimental system are described in more detail in [12, 13]. Photoillumination was obtained through the bottom p-type layer.

The doped layers in a-Si:H are nearly transparent to visible light and should be as thin as possible to minimize parasitic absorption. The minority carriers have small diffusion lengths; therefore, n-type and p-type a-Si:H are not photoactive layers. The i-layer is a region with high electric field. The light is mostly absorbed in the intrinsic i-layer, where the photo-generation occurs. The photocarriers at reverse bias voltages are swept away by the electric field in the i-layer, electrons to the n-type and holes to the p-type, and contribute mainly to drift photocurrent. Dark current increases with bias voltage as shown in **Figure 2**. It is very small in a-Si:H devices at low bias voltages and is given by thermal carrier emission from the bulk. With increased bias voltage, the injection from the doped layer increases too [15]. The signal current should be much higher than the leakage (dark) current at applied reverse bias voltage at which the electric filed, necessary to achieve full depletion inside the i-layer, collects all the photo-generated e-h pairs. At the same time, the absorbed light creates additional defects.

Defects in amorphous silicon lead to a low mobility of the charge carriers. The recombination losses of free carriers, trapping-detrapping in midgap states, and band tail states lead to photo-generated space charge in the i-layer. The space charge distribution at the p-i edge and at the n-i edge influences the internal field and screens the applied field. It is associated with the electrons and holes' capacitance in the series. In amorphous silicon, the localized states arise from their disordered nature, bond lengths, and angles between the silicon atoms. The broken or dangling bonds (DBs) arise from not-satisfied Si-Si bonds. To passivate those DBs in material is introduced the hydrogen to form the chemical bonds with the defects.

http://dx.doi.org/10.5772/intechopen.80503

The H content in the material influences the band gap values which are typically around 1.7–1.8 eV. These metastable localized states act as defect (D) states at discrete energies and as recombination centers. Dangling bonds are the main defect in a-Si:H and have defect pool model distribution and Gaussian distribution (**Figure 3**) [27]. They can be in neutral $D^0(E)$, positive $D^+(E)$, and negative $D^-(E)$ charge states and their distributions depend on light intensity and temperature. The transition $D^{+/0}$ follows acceptor statistics and $D^{0/-}$ donor statistics.

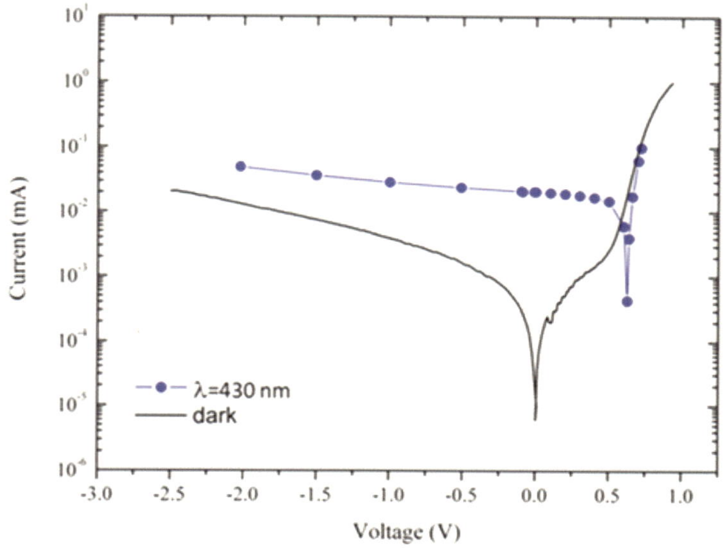

Figure 2. The p-i-n a-Si:H PD current-voltage, I-V, characteristics measured under the dark and blue LED light illumination, $\lambda = 430$ nm.

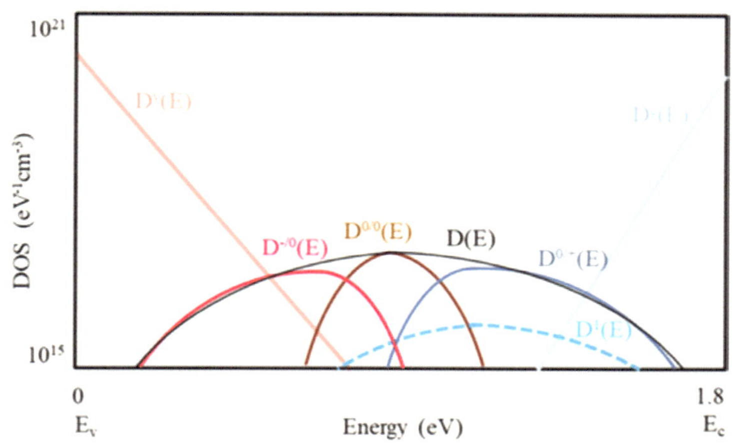

Figure 3. Scheme of band tail distribution (Dv, Dc), DOS equilibrium distribution according to defect pool model, and $D^1(E)$ acceptor-like Gaussian distribution after [27].

In a-Si:H, electrons occupying the localized states are trapped or immobile, and electrons occupying extended states are assumed to be mobile and are characterized by a "band mobility" ($\mu_n \simeq 10\,\mathrm{cm^2\,s^{-1}}$). The localized band tail states are divided from extended electron states by E_C conduction mobility edge. There are the valence mobility-edge E_V separating delocalized transport states (below E_V), where the free holes are characterized by a "band mobility" ($\mu_p \simeq 1\ \mathrm{cm^2\,s^{-1}}$), and localized traps (above E_V). The band tail states have an exponential distribution ("Urbach" tail). The conduction band tail (acceptor type) width is assumed to be $\Delta E_C = 25\,\mathrm{eV}$ and the valence band tail (donor type) width $\Delta E_V = 45-50\ \mathrm{meV}$, respectively. Hence, in a-Si:H, the mobility gap denotes the switch from small to larger mobility.

The localized state density (DOS) is so large that an electron can move from one localized site to another by hopping and the transport via these gap states is possible, but usually in numerical analysis it is neglected. The DBs act as main recombination centers. The empty gap states (trap) which interact with majority carriers via trapping-detrapping processes can be probed under sufficiently weak bias illumination level and high modulation frequency MPC method [27]. At low frequency regime, the recombination of free electrons through the recombination centers in gap distributions D(E) occupied by holes between the trap quasi-Fermi levels of electrons and holes can be probed depending on the magnitude of the capture coefficients of the recombination centers. The scheme of the DOS distribution in undoped a-Si:H, according to the defect pool model and Gaussian distribution, is shown in **Figure 3**.

Han et al. [29] have reported the most interesting feature of optical bias. Optical bias impedes deep trapping, thus enhancing electron drift. Their photocapacitance and capacitance transient measurement result indicates the band tail transport occurs in time shorter than 10 μs which is not affected by optical bias, electron trapping, and further drift following reemission from the deep trap in time longer than 1 ms.

To use the a-Si:H p-i-n photodiode as a biosensor transducer in detection of mammalian HeLa cells' chemiluminescence in our main experiment, the photodiode characterization is first done. All measurements were performed at the room temperature. LEDs (Kingbright) emitting at 430 nm for blue (B) were used in the experiment and the dc forward current through the LED was $I_F = 20$ mA. The energy of monochromatic LED light is higher than the band gap energy.

The photodiode current-voltage (I-V) characteristics measured under the dark and blue LED light illumination at $\lambda = 430$ nm are shown in **Figure 2**. Under low forward voltages, the dark current is dominated by Shockley-Read-Hall (SRH) recombination [30].

In order to obtain the information on the recombination rate in dark, the ideality factor is studied. It is well known that in a-Si:H, the ideality factor is a non-integer and decreases with temperature [30].

The recombination rate depends on the concentration of active recombination centers which include all traps between the quasi-Fermi levels for trapped charges. Second, it depends on the recombination efficacy of each of these recombination centers. These two factors are voltage dependent due to the continuous density of states in the band gap. The dark current has an exponential term and the integration term. The exponential term with ideality factor n = 2 is due to the injection of free carriers, electrons and holes, from the n and p contacts, and their recombination through a single defect level. The integration term is due to the number of defect states which act as recombination centers and are located between the quasi-Fermi levels for trapped holes, E_{fp_t}, and trapped electrons, E_{fn_t}. Their concentrations increase with increased applied voltage, as the separation of quasi-Fermi levels increases.

From dark current-voltage characteristics

$$I(V) = I_0 \exp\left(\frac{qV}{nkT}\right) \quad (1)$$

and the ideality factor defined by Deng and Wronski [31] is

$$n(V) = \left[\frac{kT}{q}\frac{d\ln(I_D)}{dV}\right]^{-1}. \quad (2)$$

The shape of the $n(V)$ curve, shown in **Figure 4**, reflects the energy distribution, as a Gaussian one, of the defect states in the *i*-layer.

The total recombination current density conducted through the device expressed by its activation energy of SRH recombination [30] shown in **Figure 5** is calculated by the expression:

$$E_a^R(V) = \frac{E_\mu - V}{2} + 3kT \quad (3)$$

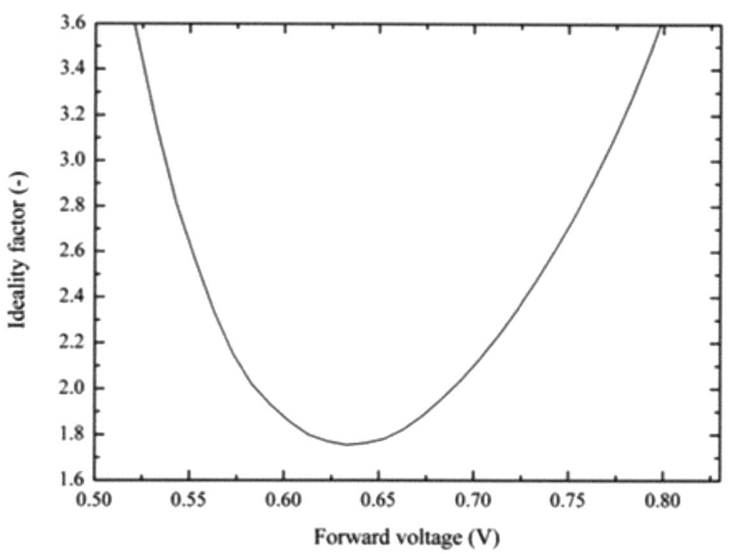

Figure 4. The voltage-dependent ideality factor, $n(V)$, as a function of voltage at room temperature for a p-i-n photodiode with an i-layer thickness of 300 nm. Calculation is done using the Deng and Wronski definition of voltage-dependent ideality factor.

where E_μ is mobility gap and V the applied voltage. Calculation is done following the Kind et al. expression for the voltage-dependent activation energy of the total recombination at various mobility gap and voltage-dependent ideality factor values shown in **Figure 4**. For comparison is given the activation energy at constant ideality factor n = 2 (the thermal ideality factor defined by Pieters et al. and used in [30]).

2.2. Photodiode capacitance

The time domain technique at low frequency is used to measure the photodiode's capacitance [32]. The measurements have been carried out on a-Si:H p-i-n cells under forward and reverse bias voltages, in dark and upon blue LED illumination and voltage pulses at 333 Hz [33].

The total charge stored in photodiode capacitor depends on the photodiode voltage as described by

$$Q = \int_0^{V_{PD}} C_{PD}\, dV \tag{4}$$

where Q is the total charge stored, C_{PD} is the photodiode capacitance as a function of voltage, and V_{PD} is the voltage across the photodiode capacitance. The current due to stored charge is

$$i(t) = \frac{dQ}{dt}. \tag{5}$$

The charge equivalent linear capacitor C_q which stores the same amount of charge as a photodiode capacitor at photodiode voltage V_{PD}, is defined as

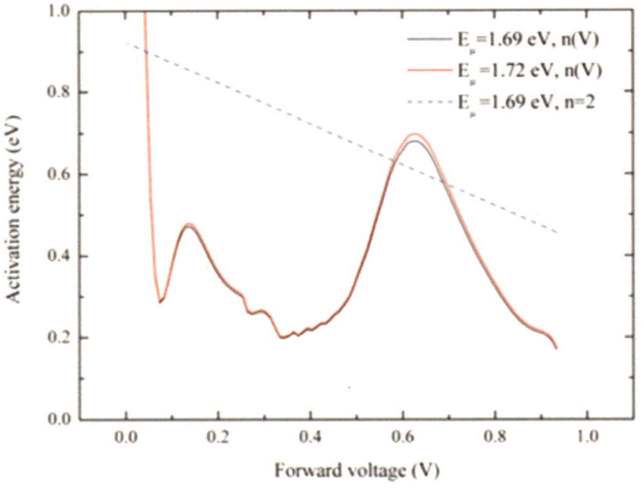

Figure 5. The activation energy as a function of voltage for an a-Si:H p-i-n photodiode with an i-layer thickness of 300 nm at room temperature.

$$C_q = \frac{Q}{V_{PD}} = \frac{\int_0^{V_{PD}} \mathrm{idt}}{V_{PD}} = \frac{\int_0^{V_{PD}} C_{PD}\,\mathrm{dv}}{V_{PD}} \tag{6}$$

The photodiode current is measured with digital storage oscilloscope (Keysight InfiniiVision 2000 X-Series Oscilloscopes) by voltage drop across the resistor. The dc bias voltage (−2 to 0.7 V) is applied and measurements are carried out. The characteristic photodiode transient response on voltage pulse is shown in **Figure 6**. The cell capacitance is calculated from total charge obtained by integration of photodiode current transient response on voltage pulse Eq. (6) and divided with the corresponding photodiode bias voltage.

The dark capacitance's dependence on photodiode voltage and capacitance under illumination with blue light is shown in **Figure 7**. It shows a quasi-linear dependence of capacitance on the voltage under illumination.

It is observed that the deep defect states of the i-layer contribute to the capacitance at various bias voltages. It is evident that around the built-in voltage, the injected charge in the dark and photo-generated charge have the same value. At higher voltages prevails the injected charge in dark. Also, the capacitance achieves the upper limit around the built-in potential. The capacitance degradation effect happens at sufficiently high forward voltages around built-in voltage (V_{bi}), where the diode injection capacitance becomes more dominant and the device responds to the voltage as a resistor.

The area under the current response curve gives the total charge (**Figure 8**) accumulated in the photodiode. In dark, at reverse bias voltages higher than 0.5 V, the changes in space charge

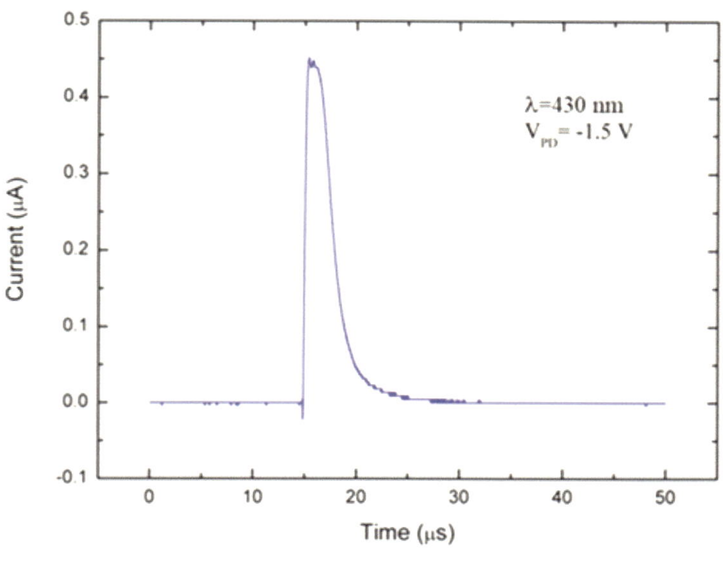

Figure 6. The transient response of a-Si:H photodiode to a square voltage pulse upon blue light illumination and reverse bias voltage $V_{PD} = -1.5$ V.

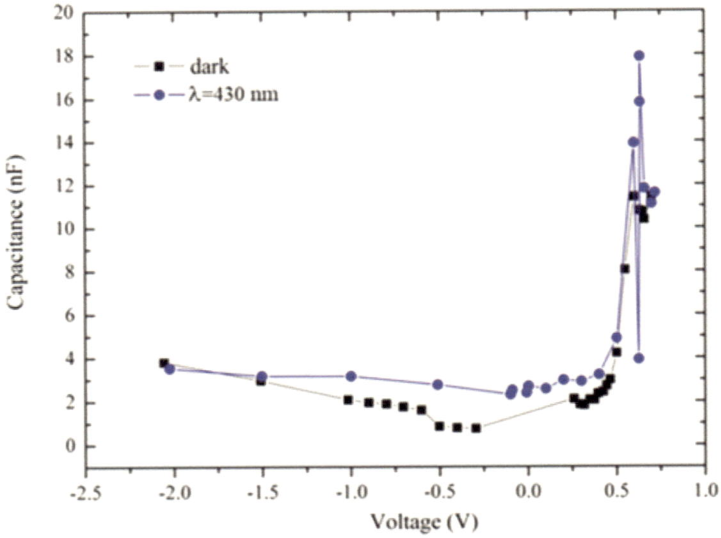

Figure 7. The a-Si:H photodiode capacitance versus bias photodiode voltages.

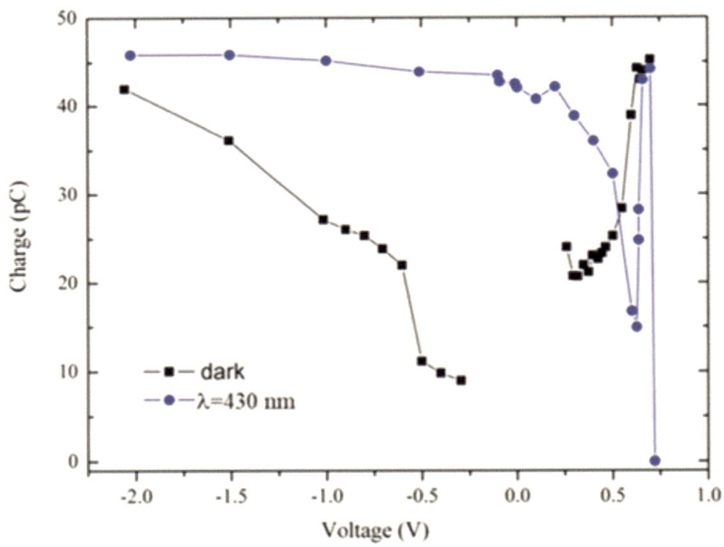

Figure 8. The a-Si:H photodiode total accumulated charge at different device voltages in dark and under blue LED light illumination (430 nm).

and local electric field in i-layer around p/i and n/i interfaces lead to the increase in total accumulated charge and consequently the capacitance increases. The increase in photo-generated charge with increased reverse bias voltage is smaller than dark charge. The proposed method can be used for further development of photodiode-integrated system and biosensors.

2.3. The blue light-induced defect creation examined with the OBMPC method

Using moderated OBMPC [11, 27, 36], we examine the light-induced defects kinetics and nature in the i-layer of a-Si:H p-i-n photodiode. Furthermore, we clarify their influence on photocurrent degradation and capacitance contribution.

The photodiode was illuminated with two blue LEDs (430 nm), a constant pump (optical bias) light and square pulse (probe) light at frequency of 333 Hz with 50% duty cycle. The intensity of the optical bias light and the pulsed probe beam was adjusted with 20-mA current through the LEDs. The illuminations were from the p-type layer side. The measurements were performed in the range from forward bias voltages of 0.7 V to reverse bias voltages of −2 V. From the measured switch-off transient response to a blue light pulse, we numerically analyze, by the generalized Foss method and general solution developed by Jeričević [33, 34], the trap and recombination localized states' energy distribution in the energy gap. The number of components, not known in advance, in multiexponential decay of measured switch-off transient response is determined by its best fit with numerically modeled transient response.

The photo-generated electron-hole pairs upon blue light illumination are nonuniformly generated near the front surface in the vicinity of the p^+/i interface. The photo-generated free carrier densities, electrons, and holes, have dc and time-dependent pulsed components.

The holes' contribution to the transient photocurrent is small, due to their trapping near the p^+/i interface where arises the space charge density or their movement into the front contact [11].

We observe a short time delay of transient photocurrent ascribed to trapping and release interaction of free carriers with shallow band gap localized states. The transient photocurrent decay in tail-like form, dependent on applied voltage, often happens due to deep trapping. It is dependent on the time that an electron spent in discrete localized states $N(E_i)$ at E_i energy levels (capture and release), as described in [11].

$$\tau = v_0^{-1} e^{(E_i/kT)}. \tag{7}$$

Based on the MPC theory described in [36], the band gap energy is divided into three energy ranges. The energies from which electrons (holes) can be trapped and released to the conduction (valence) band, E, are above, $E > E_{tn}$ (below, $E < E_{tp}$) quasi-Fermi level for trapped electrons, E_{tn} (trapped holes, E_{tp}), and recombination states between quasi-Fermi level for trapped electrons, E_{tn} and holes, E_t , $E_{tp} < E < E_{tn}$.

The position of quasi-Fermi level for trapped electrons, E_{tn} [35, 36], is determined from the measured dc photocurrent at applied bias voltage and under constant illumination from Eq. (8) in [36]. At E_{tn} and $E_{tp,}$ the occupation function, $f_{dc,}$ of gap states changes from 1 to 0 in two steps. The dc generation rate characteristic time response (Eq. (1) in [36]), extracted from measured photocurrent transient response is compared with the characteristic time of the experiment, taking in to account the characteristic capture frequency ω_c (Eq. (2) in [36]) to determine the high or low frequency regime of the experiment.

From calculated values, in our experiment, the chosen frequency falls in the low-frequency regime. In this regime, only the defects around the Fermi level can be probed. The requirement that Fermi level of free electrons coincides with quasi-Fermi level of trapped electron will be satisfied.

The measured a-Si:H p-i-n photodiode switch-off photocurrent transient response on blue probe light at blue bias light and 0 V bias voltage on 10 kΩ load resistor, the calculated transient response, and difference between them are shown in **Figure 9**. The two exponential functions, as in **Figure 9**, are present in all the cases of applied bias voltage.

Figure 10 shows the numerically extracted energies of localized states from measured photocurrent transient response. The weighting factor (pre-exponential factor) of localized states is shown in **Figure 11**. The weighting factors corresponding to the deeper gap states (E_2) are higher than those of the shallower (E_1) states for all voltages below the built-in voltage. With increasing forward bias voltage, there is an increase in weighting factor corresponding to energy E_1 and decrease in those of energy E_2. The energy levels E_1 and E_2 shift toward deeper energy levels for moderate forward voltages below the built-in voltage. At high forward voltages, both shift toward shallower values. This is in agreement with [27], where the capture coefficients of the midgap states were higher than those of the shallow localized states. Also, these results confirm the capacitance upper limit described above (**Figures 7** and **8**).

2.4. a-Si:H p-i-n photodiode as a transducer in biosensors

By definition of Mehrotra, biosensors are analytical devices that convert a biological response into an electrical signal [37]. They have many applications in medical diagnostics, pharmaceutical, food, beverage, agricultural, environmental, and biotechnological industries. Two main components of biosensors are the bioreceptor and transducer [38, 39]. Bioreceptor is a

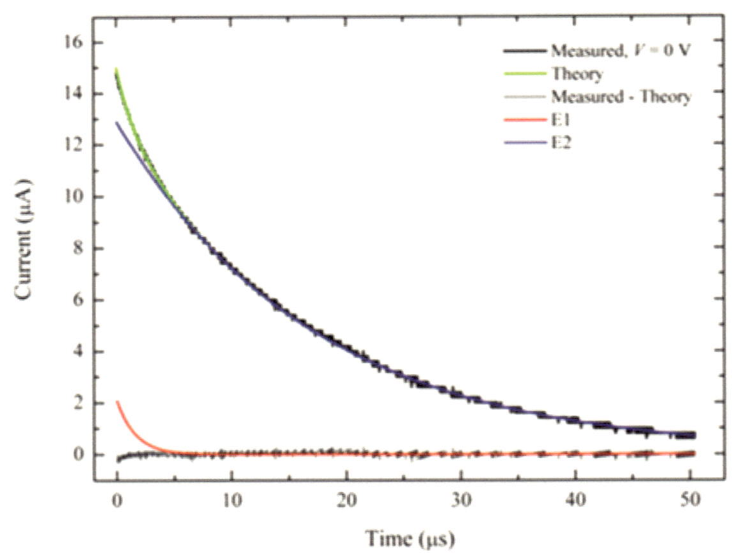

Figure 9. The measured a-Si:H p-i-n photodiode switch-off photocurrent transient response on blue probe light at blue bias light and 0 V bias voltage on 10 kΩ load resistor, the numerically reconstructed transient response (theory), and difference between them.

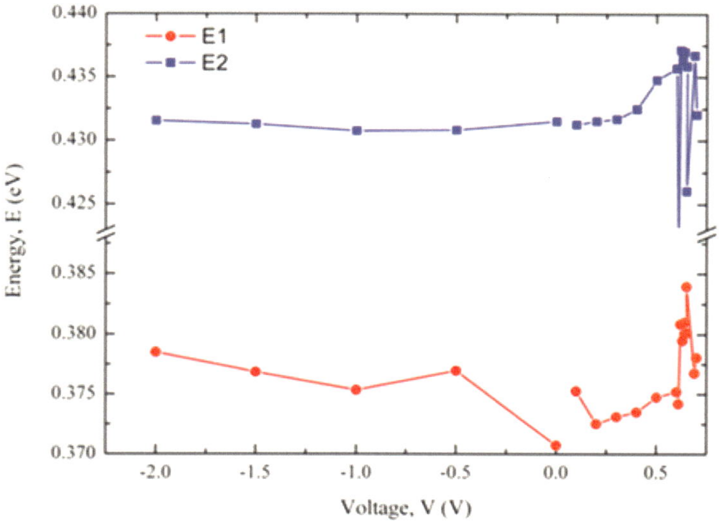

Figure 10. The energies of localized states extracted from measured photocurrent transient response of a-Si:H p-i-n PD on blue probe light at blue bias light at applied voltages Vappl. = −2 to 0.7 V.

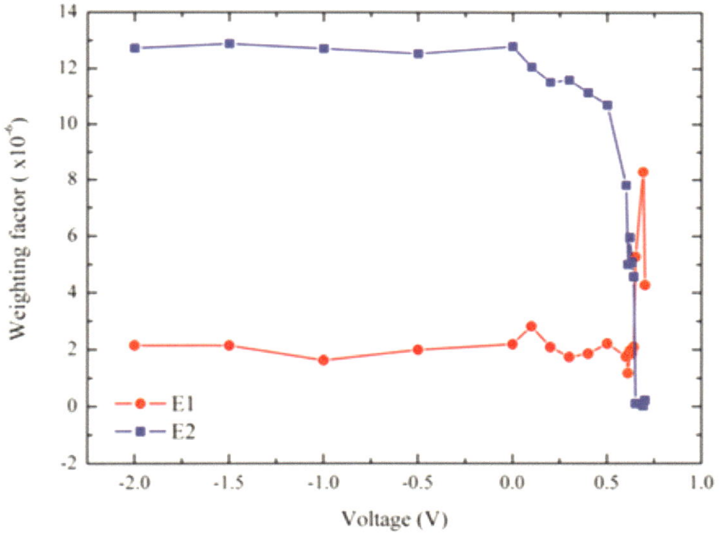

Figure 11. Weighting factor of localized states extracted from measured photocurrent transient response of a-Si:H p-i-n PD on blue probe light at blue bias light at applied voltages Vappl. = −2 to 0.7 V.

part that recognizes the analyte of interest, while biotransducer is a physicochemical detector that converts the bioreceptor-analyte complex into a measurable signal. As the name says, a bioreceptor is a biological molecule like enzymes, antibodies, and nucleic acid, but it can also be a tissue, organelle, or microorganism, while the biotransducer's measurable signal may be viscosity, mass, temperature, electrical current, electrical potential, impedance, conductance, electromagnetic field, electromagnetic radiation, or visible light. Biosensors can be label-free or label based which depends on their detection system [40].

Biosensing elements can be described as follows:

Enzymes: protein molecule which acts as a catalyst in chemical reactions. They can be mobilized on transducers by gel entrapment technology, covalent binding, or physical adsorption.

Microbes: they are capable of transforming analytes to specific products which can be monitored by transducer.

Organelle: more specific for analysis.

Antibodies: highly selective to antigens and can be attached to matrix surface of transducer.

Nucleic acids: are DNA and RNA molecules which can be hybridized with other nucleic acids, so it can be a good sensing element for metabolic disorders, infection disease, cancer, and genetic disorders.

Aptamers: those are single-stranded DNA or RNA molecules and can be specific against amino acids, proteins, and other molecules by adopting specific and stable secondary structures against mentioned analytes.

Biosensors can be classified as electrochemical, mass dependent, optical, radiation sensitive [39], or piezoelectric based on their transduction principle. Based on the detected analyte, they can be immunosensors, aptasensors, genosensors, or enzymatic biosensors.

Optical biosensors have light as the output transducer signal. Light is generated by optical diffraction and electrochemiluminescence as main mechanisms for light production [41]. Bioluminescence is a process in which biomolecules absorb light, from the excitation source and enter into excited state, then fall down to the ground state and emit light as fluorescence or phosphorescence. Chemiluminescence is a type of luminescence when the light is emitted by chemical reaction. If the chemical reaction is catalyzed by an enzyme, it is called bioluminescence [42].

Regard, their above described characteristics, the a-Si:H photodiodes have become driving force in the scientific community for detection of tumor cells. For in vitro testing of HeLa cells, it is important to note that:

1. Cells are standardly grown in complemented Dulbecco's Modified Eagle Medium (DMEM) with fetal bovine serum addition.
2. For counting, cells are removed from the surface plate by use of enzyme trypsin.
3. All the components (cells, DMEM, and trypsin) absorb blue light.

DMEM (Dulbecco's Modified Eagle Medium) has been proposed for culturing normal and tumor cells. Constituents of the medium are high level of glucose, essential minerals, amino acid, and vitamins. Alone, it does not function for cell culturing; so, it must be complemented with fetal bovine serum, antibiotics, and L-glutamine. The components of the complemented medium DMEM that absorb blue light are riboflavin (vitamin B12), hemoglobin, and bilirubin

[43–45]. From previous works, it is known that bilirubin and riboflavin decompose under exposure of blue light [43, 46]. There are numerous factors that influence photodegradation, like radiation source, intensity, wavelength, pH, buffer, solvent polarity, and viscosity [43]. The influence of blue light on (a) complemented DMEM medium and (b) HeLa cells can be monitored by amorphous silicon (a-Si:H) photodiodes.

The photodiode's (Department of Information Engineering, Electronics and Telecommunications, Sapienza Università di Roma, Italy) p-doped/intrinsic/n-doped junction of a:Si-H layers were deposited on 50 × 50 × 1.1 mm^3 glass substrate and arranged in 5 × 6 array. The a-Si:H layers were deposited by plasma-enhanced chemical vapor deposition (PECVD) in a three-chamber high-vacuum system. The bottom electrode is a 180–nm-thick indium tin oxide (ITO) layer. The top metal electrode is a three-metal-layer stack (30-nm-thick Cr/150-nm-thick Al/30-nm-thick Cr). The area of each photodiode is 2 × 2 mm^2. Further details on the photodiode array fabrication can be found in [47].

The a-Si:H photodiode illuminated with blue LED light (RGB LED Lamp Kingbright emitting at 430 nm) is placed in a dark metallic box. The LED current was fixed at 20 mA to provide constant illumination. A reverse bias voltage equal to 2 V was applied to the photodiode. The measurements for calibration were performed at room temperature for 1 h. Before starting the assay, the 3-ml solutions containing the appropriate concentration of HeLa cells in DMEM and DMEM, respectively, are introduced with pipette in a plastic well posted on a photodiode surface. The box is then closed to minimize room light interference. The measurements are performed at 2-V reverse bias. The photodiode current and voltage are monitored for 1 h using the Keysight BenchVue software. The photodiode is connected in series with a load resistor, R_L, of 10 kΩ, voltage source (Agilent Technologies E3631A DC voltage source), and digital multimeters, DMMs (Agilent Technologies 34450A meter). Before starting the assay, the a-Si:H photodiode is illuminated with white light to neutralize the defects induced with previous blue light illumination and to reverse the process of decreasing of photoconductivity.

The significant changes in current are observed in first 20 minutes. The current characteristic transients corresponding to blue LED-induced HeLa cells' chemiluminescence detected by a-Si:H p-i-n photodiode are shown in **Figure 12**.

It can be deduced from the **Figure 12** that the photocurrent initially decreases due to creation of two types of defects under blue light illumination. The measured photocurrent (a) when 3 mL of complemented DMEM and trypsin are placed in plastic well has faster exponential decay than the photodiode in first 2 min. The decrease in photocurrent can be attributed to absorption of blue light in the DMEM solution and in the photodiode. After 20 min, the photocurrent decreases which can be attributed to the decomposition of riboflavin (not presented here). It is also known from the literature that bilirubin and riboflavin obey first-order decomposition kinetics when they are exposed to blue light; and although the kinetic coefficient for riboflavin is 10 times greater than for bilirubin, it can be speculated that riboflavin decomposes in our experiment [46, 48].

Effects of visible spectra on live organisms have been studied for different approaches. Light can induce photochemical reactions in living cells and can have benefits in treatment of some

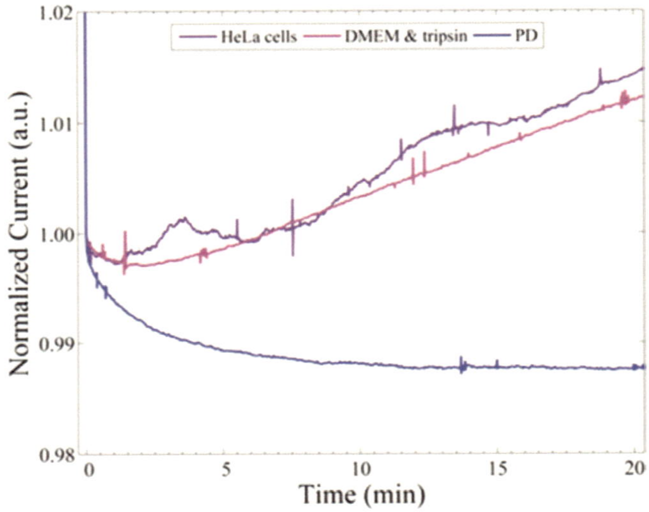

Figure 12. Normalized measured a-Si:H p-i-n photodiode, PD photocurrent versus time, with HeLa cells, and DMEM and trypsin, respectively, in plastic well on PD surface.

diseases, that is, psoriasis and neonatal hyperbilirubinemia [49, 50]. It can modulate the endocrine system and accelerate the maturation of ovaries in young rats [51].

Blue light can influence skin-keratinocytes exerting antiproliferative effect and inducing differentiation; so, it can have therapeutic effects for hyperproliferative skin conditions [52]. Effects of blue light on human health are very beneficial because it can inhibit the growth of tumors, killing bacterial spores or inactivate microorganisms [53–55]. There are a number of chromophores inside cells that absorb blue light like riboflavin, flavin proteins, iron-sulfur proteins, cytochromes, etc. a-Si:H p-i-n photodiode can be a good detector with high sensitivity, good spectral responsivity, and small reflectance for blue light, for measuring low light intensity in visible spectrum (430–780 nm). So, (b) in experiment with HeLa cells under blue light illumination the low intensity light, which is product of chemiluminescence inside cells, can be detected. According to this experiment, HeLa cells under illumination with blue light exert dramatic changes in their metabolic activity. It is well known that blue light can induce hydrogen peroxide production in mammalian cells, and release nitrogen oxide from nitrosylated proteins [52, 53]. In tumor HeLa cells, nitric oxide modulates a number of biological processes which can be witnessed by increase in NO-synthetase levels [56]. Also, it is well known that nitric oxide and hydrogen peroxide can react and release light from chemiluminescence reaction producing the toxic reactive oxygen species singlet oxygen [57]. Singlet oxygen can induce serious damage in cells and could kill 43% of tumor cells in 1 h in our experiments. So, the blue light has two effects on tumor HeLa cells: inducing chemiluminescence and killing tumor cells. Chemiluminescence can be detected by a-Si:H photodiode and that chemiluminescence reaction rate versus time sequence obeys the exponential decay. **Figure 13** shows the difference in photocurrent of HeLa cells and DMEM and trypsin. It can be deduced that absorption of HeLa cells can be separated from complemented DMEM medium. It can be

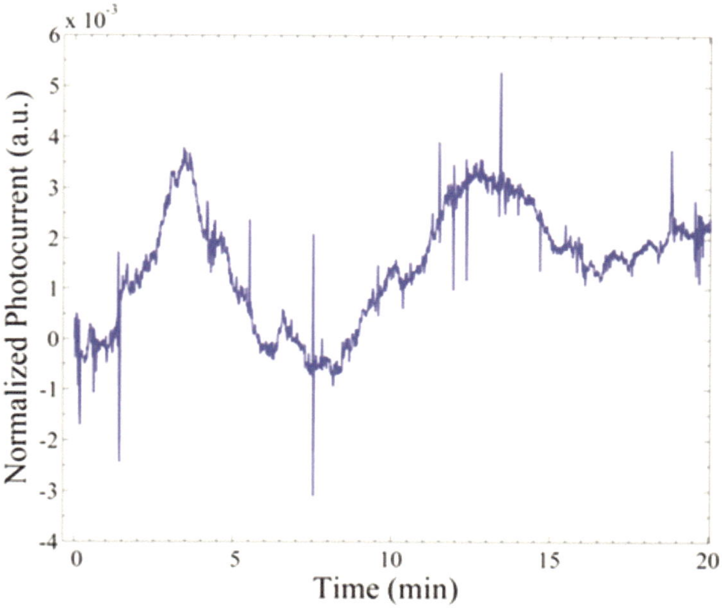

Figure 13. The normalized photocurrent difference of HeLa cells and DMEM and trypsin.

concluded that HeLa cells produce chemiluminescence radiation in the visible part of spectrum, while in the DMEM solution, this is not observed.

3. Conclusions

We performed an experiment on mammalian cells' chemiluminescence detection based on the phenomenon that under illumination of two-beam, reverse-biased a-Si:H photodiode current exceed expected primary photocurrent. The native and metastable defects in a-si:H p-i-n phtodiodes activated in this phenomenon are first characterized with simultaneous blue light pulse and constant blue light illumination at low frequency. From a transient response, the photocapacitance is analyzed. Finally, the HeLa cells' chemiluminescence reaction measurement method is done. It can be concluded that a-Si:H photodiodes can be good transducers in optical biosensors for detecting tumor cells and chemiluminescence reaction inside cells.

Acknowledgements

The authors thank Prof. Domenico Caputo and Prof. Giampiero de Cesare at University of Rome "La Sapienza", Italy and group from process line at Solar cells d.o.o. Split Croatia for provision of samples used in this work.

Conflict of interest

The authors declare that there are no conflict of interests regarding the publication of this paper.

Author details

Vera Gradišnik[1]* and Darko Gumbarević[2]

*Address all correspondence to: vera.gradisnik@uniri.hr

1 Faculty of Engineering, University of Rijeka, Rijeka, Croatia

2 Department of Biotechnology, University of Rijeka, Rijeka, Croatia

References

[1] Robbins H, Sumitomo K, Tsujimura N, Kamei T. Integrated thin film Si fluorescence sensor coupled with a GaN microLED for microfluidic point-of-care testing. Journal of Micromechanics and Microengineering. 2018;**28**:024001. DOI: 10.1088/1361-6439/aa9e6d

[2] Santos DR, Soares RRG, Chu V, Conde JP. Performance of hydrogenated amorphous silicon thin film photosensors at Ultra-Low light levels: Towards attomole sensitivities in Lab-on-Chip biosensing applications. IEEE Sensors Journal. 2017;**17**:6895-6903. DOI: 10.1109/JSEN. 2017.2751253

[3] Kamei T, Paegel BM, Scherer JR, Skelley AM, Street RA, Mathies RA. Integrated hydrogenated amorphous Si photodiode detector for microfluidic bioanalytical devices. Analytical Chemistry. 2003;**75**:5300-5305. DOI: 10.1021/ac0301550

[4] Feenstra KF, Schropp REI, Van der Weg WF. Deposition of amorphous silicon films by hot-wire chemical vapor deposition. Journal of Applied Physics. 1999;**85**:6843-6852. DOI: 10.1063/1.370202

[5] Moraes D, Anelli G, Despeisse M, Disertori G, Garrigos A, Jarron P, et al. A novel low noise hydrogenated amorphous silicon pixel detector. Journal of Non-Crystalline Solids. 2004;**338-340**:729-731. DOI: 10.1109/NSSMIC.2005.1596579

[6] Yoshida N, Shimizu Y, Honda T, Yokoi T, Nonomura S. A study of absorption coefficient spectra in a-Si:H films near the transition from amorphous to crystalline phase measured by resonant photothermal bending spectroscopy. 2008;**354**:2164-2166. DOI: 10.1016/j.jnoncrysol.2007.10.039

[7] Fossum ER. Active pixel sensors: Are CCDs dinosaurs? In: Proceedings of the SPIE 1900 Charge-Coupled Devices and Solid State Optical Sensors III; 12 July 1993. San Jose, CA, US: SPIE; 1900. pp. 2-14. DOI: 10.1117/12.148585

[8] Izadi MH, Tousignant O, Feuto Mokam M, Karim SK. An a-Si active pixel sensor (APS) array for medical X-ray imaging. IEEE Transactions on Electron Devices. 2010;**57**: 3020-3026. DOI: 10.1109/TED.2010.2069010

[9] Karim KS, Nathan A. Readout circuit in active pixel sensors in amorphous silicon technology. IEEE Electron Device Letters. 2001;**22**:469-471. DOI: 10.1109/55.954914

[10] Karim KS, Nathan A. Amorphous silicon active pixel sensor readout circuit for digital imaging. IEEE Transactions on Electron Devices. 2003;**50**:200-208. DOI: 10.1109/TED.2002.806968

[11] Shen DS, Wagner S. Transient photocurrent in hydrogenated amorphous silicon and implications for photodetector devices. Journal of Applied Physics. 1996;**79**:794-801. DOI: 10.1063/1.360827

[12] Gradišnik V, Pavlović M, Pivac B, Zulim I. Study of the color detection of a-Si:H by transient response in the visible range. IEEE Transactions on Electron Devices. 2002;**49**: 550-556. DOI: 10.1109/16.992861

[13] Gradišnik V, Pavlović M, Pivac B, Zulim I. Transient response times of a-Si:H p-i-n color detector. IEEE Transactions on Electron Devices. 2006;**53**:2485-2491. DOI: 10.1109/TED.2006.882265

[14] Fuhs W. Recombination and transport through localized states in hydrogenated amorpohus and microcrystalline silicon. Journal of Non-Crystalline Solids. 2008;**354**:2067-2078. DOI: 10.1016/j.jnoncrysol.2007.09.008

[15] Murthy VR, Dutta V. Underlying reverse current mechanisms in a-Si:H p+–i-n+ solar cell and compact SPICE modelling. Journal of Non-Crystalline Solids. 2008;**354**:3780-3784. DOI: 10.1016/j.jnoncrysol.2008.03.041

[16] Mahmood SA, Kabir MZ. Modeling of transient and steady-state dark current in amorphous silicon p-i-n photodiodes. Current Applied Physics. 2009;**9**:1393-1396. DOI: 10.1016/j.cap.2009.03.011

[17] Dhariwal SR, Rajvanshi S. Theory of amorphous silicon solar cell (a): Numerical analysis. Solar Energy Materials & Solar Cells. 2003;**79**:199-213. DOI: 10.1016/S0927-0248(02)00414-2

[18] Dhariwal SR, Rajvanshi S. Theory of amorphous silicon solar cell (b): A five layer analytical model. Solar Energy Materials & Solar Cells. 2003;**79**:215-233. DOI: 10.1016/S0927-0248(02)00415-4

[19] Dhariwal SR, Smirty M. On the sensitivity of open-circuit voltage and fill factor on dangling bond density and Fermi level position in amorphous silicon p-i-n solar cell. Solar Energy Materials & Solar Cells. 2006;**90**:1254-1272. DOI: 10.1016/j.solmat.2005.08.001

[20] Kopprio L, Longeaud C, Schmidt J. Obtainment of the density of states in the band tails of hydrogenated amorphous silicon. Journal of Applied Physics. 2017;**122**:085702. DOI: 10.1063/1.4999626

[21] Vanecek M, Kočka J, Poruba A, Fejfar A. Direct measurement of the deep defect density in thin amorphous silicon films with the "absolute" constant photocurrent method. Journal of Applied Physics. 1995;**78**:6203-6210. DOI: 10.1063/1.360566

[22] Melskens J, Smets AHM, Schouten M, Eijt SWH, Schut H, Zeman M. New insights in the nanostructure and defect states of hydrogenated amorphous silicon obtained by annealing. IEEE Journal of Photovoltaics. 2013;**3**:65-71. DOI: 10.1109/JPHOTOV.2012.2226870

[23] Melskens J, Schouten M, Santbergen R, Fischer M, Vasudevan R, vanderVlies DJ, et al. In situ manipulation of the subgap states in hydrogenated amorphous silicon monitored by advanced application of Fourier transform photocurrent spectroscopy. Solar Energy Materials & Solar Cells. 2014;**129**:70-81. DOI: 10.1016/j.solmat.2014.03.022

[24] Jiao L, Liu H, Semoushikina S, Lee Y, Wronski CR. Initial, rapid light-induced changes in hydrogenated amorphous silicon materials and solar cell structures: The effects of charged defects. Applied Physics Letters. 1996;**69**:3713. DOI: 10.1063/1.117198

[25] Melskens J, Schouten M, Mannheim A, Vullers AS, Mohammadian Y, Eijt SWH, et al. The nature and the kinetics of light-induced defect creation in hydrogenated amorphous silicon films and solar cells IEEE. Journal of Photovoltaics. 2014;**6**:1331-1336. DOI: 10.1109/JPHOTOV.2014.2349655

[26] Zollondz J-H, Reynolds S, Main C, Smirnov V, Zrinscak I. The influence of defects on response speed of high gain two-beam photogating in a-Si:H PIN structures. Journal of Non-Crystalline Solids. 2002;**299-302**:594-598. DOI: 10.1016/S0022-3093(01)01204-2

[27] Pomoni M, Kounavis P. Determination of trapping–detrapping events, recombination processes and gap-state parameters by modulated photocurrent measurements on amorphous silicon. Philosophical Magazine. 2015;**94**:2447-2471. DOI: 10.1080/14786435.2014.914262

[28] Gradišnik V, Jeričević Ž. The a-Si:H device characteristics degradation upon the light induced defects. In: Book of Abstracts of the 3rd Euroregional Workshop on Photovoltaics & Nanophotonics 2016 (EUROREG PV 2016); 21-23 September 2016; Lubljana. Slovenia: LPVO Faculty of Electrical Engineering; 2016. p. 21

[29] Han D, Melcher DC, Schiff EA. Optical-bias effects in electron-drift measurements and defect relaxation in a-Si:H. Physical Review B. 1993;**48**:8658-8666. DOI: 10.1103/PhysRevB.48.8658

[30] Kind R, van Swaaij RAACMM, Rubinelli FA, Solntsev S Zeman M. Thermal ideality factor of hydrogenated amorphous silicon p-i-n solar cells. Journal of Applied Physics. 2011;**110**:104512. DOI: 10.1063/1.3662924

[31] Deng J, Wronski CR. Carrier recombination and differential diode quality factors in the dark forward bias current-voltage characteristics of aSi:H solar cells. Journal of Applied Physics. 2005;**98**:024509. DOI: 1063/1.1990267

[32] Kumar RA, Suresh MS, Nagaraju J. Time domain technique to measure solar cell capacitance. Review of Scientific Instruments. 2003;**74**:3516-3519. DOI: 10.1063/1.1582391

[33] Čović M, Gradišnik V, Jeričević Ž. The investigation of influence of localized states on a-Si:H p-i-n photodiode transient response to blue light impulse with blue light optical bias. In: Proceedings of the 39th International Convention on Information and Communication Technology, Electronics and Microelectronics (MIPRO 2016); 30 May-3 June 2016; Opatija. Croatia: IEEE; 2016. pp. 24-27. DOI: 10.1109/MIPRO.2016.7522104

[34] Jericevic Z. Method for Fitting a Sum of Exponentials to Experimental Data by Linearization Using Numerical Integration Approximation, and Its Application to Well Log Data. [US Patent] USP #7,088,097; 2006

[35] Kounavis P. Changes in the trapping and recombination process of hydrogenated amorphous silicon in the Staebler–Wronski effect. Journal of Applied Physics. 1995;**77**: 3872-3878. DOI: 10.1063/1.358565

[36] Kleider J-P, Longeaud C, Gueunier M-E. The modulated photocurrent technique: A powerful tool to investigate band gap states in silicon based thin films. Physica Status Solidi (C). 2004;**1**:1208-1226. DOI: 10.1002/pssc.200304322

[37] Mehrotra P. Biosensors and their applications—A review. Journal of Oral Biology and Craniofacial Research. 2016;**6**:153-159. DOI: 10.1016/j.jobcr.2015.12.002

[38] Darsanaki RK, Azizzadeh A, Nourbakhsh M, Raeisi G, Aliabadi MA. Biosensors: Functions and applications. Journal of Biology and Today's World. 2013;**2**:53-61. DOI: 10.15412/J. JBTW.01020105

[39] Ali J, Najeeb J, Ali MA, Aslam MF, Raza A. Biosensors: Their fundamentals, designs, types and most recent impactful applications: A review. Journal of Biosensors and Bioelectronics. 2017;**8**:9. DOI: 10.4172/2155-6210.1000235

[40] Vigneshvar S, Sudhakumari CC, Senthilkumaran B, Prakash H. Recent advances in biosensor technology for potential applications—An overview. Frontiers in Biotechnology and Bioengenieering. 2016;**4**:11. DOI: 10.3389/fbice.2016.00011

[41] Duval D, Lechuga LM. Breakthroughs in Photonics 2012: 2012 Breakthroughs in Lab-on-a-Chip and Optical Biosensors. IEEE Photonics Journal. 2013;**5**. DOI: 10.1109/ JPHOT.2013.2250943

[42] Nabi A. Instrumentation for chemiluminescence and bioluminescence assays (Part-I). Journal Chemical Society of Pakistan. 1991;**13**:90-95

[43] Sheraz MA, Kazi SH, Ahmed S, Anwar Z, Ahmad I. Photo, thermal and chemical degradation of riboflavin. Beilstein Journal of Organic Chemistry. 2014;**10**:1999-2012. DOI: 10.3762/bjoc.10.208

[44] Horecher BL. The absorption spectra of hemoglobin and its derivatives in the visible and near infra-red regions. Journal of Biological Chemistry. 1943;**148**:173-183

[45] McDonagh AF, Palma LA, Lightner DA. Blue Light and Bilirubine Excretion. Science. 1980;**208**:145-151. DOI: 10.1126/science.7361112

[46] Ahmad HB, Ahmad S, Shad MA, Hussain M. Kinetic measurement of photodecomposition of bilirubin. Asian Journal of Chemistry. 2013;**25**:7945-7948. DOI: 10.14233/ajchem.2013.14749

[47] Constantini F, Sberna C, Petrucci G, Reverberi M, Domenici F, Fanelli C, et al. Aptamer-based sandwich assay for on chip detection of Ochratoxin A by an array of amorphous silicon photosensors. Sensors and Actuators B: Chemical. 2016;**230**:31-39. DOI: 10.1016/j.snb.2016.02.036

[48] Sattar A, de Man JM, Alexander JC. Light-Induced degradation of Vitamins I. Kinetic studies of riboflavin decomposition in solution. Canadian Institute of Food Science and Technology journal. 1977;**10**:61-64. DOI: 10.1016/S0315-5463(77)73439-X

[49] Ferguson J. A controlled study of ultraviolet A sunbed treatment of psoriasis. British Journal of Dermatology. 2000;**143**:919-920. DOI: 10.1046/j.1365-2133.2000.03820.x

[50] Salih FM. Can sunlight replace phototherapy units in the treatment of neonatal jaundice? An in vitro study. Photodermatology, Photoimmunology & Photomedicine. 2001;**17**(6): 272-277. DOI: 10.1111/j.1600-0781.2001.170605.x

[51] Wurtman RJ. The effects of light on the human body. American Journal of Sciences. 1975;**233**(1):69-77. DOI: 10.1038/scientificamerican0775-68

[52] Liebmann J, Born M, Kolb-Bachofen V. Blue-light irradiation regulates proliferation and differentiation in human skin cells. Journal of Investigative Dermatology. 2010;**130**(1): 259-269. DOI: 10.1038/jid.2009.194

[53] Sparsa A, Faucher K, Sol V, Durox H, Boulinguez S, Doffoel-Hantz V, et al. Blue light is phototoxic for B16f10 murine melanoma and bovine endothelial cell lines by direct cytocidal effect. Anticancer Research. 2010;**30**(1):143-148

[54] Denis TGS, Dai T, Hamblin MR. Killing bacterial spores with blue light: When innate resistance meets the powerof light. Photochemistry and Photobiology. 2013;**89**(1):2-4. DOI: 10.1111/j.1751-1097.2012.01233.x

[55] Liang J-Y, Yuann J-MP, Cheng CW, Jian HL, Lin C-C, Yu Chen L. Blue light induced free radicals from riboflavin on *E. coli* DNA damage. Journal of Photochemistry and Photobiology B: Biology. 2013;**119**:60-64

[56] Choudhari SK, Chaudhary M, Badge S, Gadbail AR, Joshi V. Nitric oxide and cancer: A review. World Journal of Surgical Oncology. 2013;**11**:1-11. DOI: 10.1186/1477-7819-11-118

[57] Noronha-Dutra AA, Epperlein MM, Woolf N. Reaction of nitric oxide with hydrogen peroxide to produce potentially cytotoxic singlet oxygen as a model for nitric oxide-mediated killing. FEBS Letters. 1993;**321**:59-62. DOI: 10.1016/0014-5793(93)80621-Z

Section 3

Technology and Applications

Chapter 6

Single-Pixel Imaging Using Photodiodes

Yessenia Jauregui-Sánchez, Pere Clemente,
Pedro Latorre-Carmona, Jesús Lancis and
Enrique Tajahuerce

Additional information is available at the end of the chapter

http://dx.doi.org/10.5772/intechopen.79734

Abstract

Single-pixel cameras (SPCs) have been successfully used in different imaging applications during the last decade. In these techniques, the scene is illuminated with a sequence of microstructured light patterns codified onto a programmable spatial light modulator. The light coming from the scene is collected by a bucket detector, such as a photodiode. The image is recovered computationally from the photodiode electric signal. In this context, the signal quality is of capital value. One factor that degrades the signal quality is the noise, in particular, the photocurrent, the dark-current, and the thermal noise sources. In this chapter, we develop a numerical model of a SPC based on a photodiode, which considers the characteristics of the incident light, as well as the photodiode specifications. This model includes the abovementioned noise sources and infers the signal-to-noise ratio (SNR) of the SPCs in different contexts. In particular, we study the SNR as a function of the optical power of the incident light, the wavelength, and the photodiode temperature. The results of the model are compared with those obtained experimentally with a SPC.

Keywords: single-pixel cameras, structured illumination, photodiodes, signal-to-noise ratio, photocurrent, dark current, noise in imaging systems

1. Introduction

Computational imaging with a single-pixel camera (SPC), or single-pixel imaging (SPI), is a remarkable alternative to conventional imaging [1]. SPCs are based on sampling the scene with a sequence of microstructured light patterns codified onto a programmable spatial light modulator (SLM), while the intensity of the light coming from the object is measured by a detector without spatial resolution. The image is computationally recovered from the fluctuations of the electric signal provided by the detector. Thus, the quality of this temporal signal is a key factor in order to recover a high quality image.

The simplicity of the detection stage in SPI is one of the main advantages of the technique. It can be exploited to use very sensitive light sensors in low light level applications [2, 3]. It is also useful in order to measure the spatial distribution of different parameters such as the spectral content [2, 4] or the polarization state [5] of the light coming from the objects. Besides, SPI has shown robustness to the presence of scattering media [6, 7]. Moreover, the SPC can be an interesting choice for imaging using light with a spectrum beyond the visible region, such as in the infrared (IR) and terahertz spectral regions. Furthermore, SPI techniques are very well suited for the application of compressive sampling (CS, also referred to as compressive sensing) methods, which noticeably reduce the measurement time, the bottleneck of this technique [8, 9].

Among the different possible detectors, photodiodes are the most common sensors in general single-pixel imaging applications. In this chapter, we develop a numerical model of a single-pixel camera based on a photodiode, which considers the characteristics of the incident light, as well as the photodiode specifications [10]. Our model takes into account the photocurrent, the dark current, the photocurrent shot noise, the dark-current shot noise, and the Johnson-Nyquist (thermal) noise. In particular, we study the signal-to-noise ratio (SNR) as a function of the optical power level and the wavelength of the incident light, as well as the photodiode temperature. We restrict our study to silicon (Si) and indium-gallium-arsenide (InGaAs) photodiodes.

In the following sections, first, we describe in detail the single-pixel cameras. Second, we review the properties of the electrical signal provided by photodiodes based on Si and InGaAs materials. Third, we present a numerical model of the single-pixel camera. Next, we apply this model to study the SNR of a single-pixel camera in different contexts. After that, we compare some of these numerical results with those experimentally obtained in the laboratory. Finally, we emphasize the main conclusions.

2. The single-pixel camera

The SPC is able to provide images of a scene with a bucket detector, such as a photodiode, by using light-structured illumination. A schematic representation of the optical system is shown in **Figure 1**. A set of microstructured light patterns is codified onto a programmable spatial light modulator (SLM) and sequentially projected onto the scene to be analyzed. The light reflected (or transmitted) by the scene is collected by a lens and focused onto a photodiode. The photodiode provides us with an electrical current proportional to the integrated light intensity, which is digitized by a data acquisition system (DAQ). The photodiode signal represents the dot product between each microstructured light pattern and the scene. The image is retrieved from the photodiode signal and the microstructured light patterns using computational algorithms.

The ideas on which SPCs are based were proposed by Golay in 1949, for spectroscopy applications [11], and by Decker in 1970, for imaging [12]. However, the first efficient SPC was created in 2006 [13], by using a programmable SLM. The most common types of SLMs are the liquid crystal spatial light modulator (LC-SLM) and the digital micromirror device (DMD) [14]. In general, DMDs are more used than LC displays, in SPI applications, except when phase

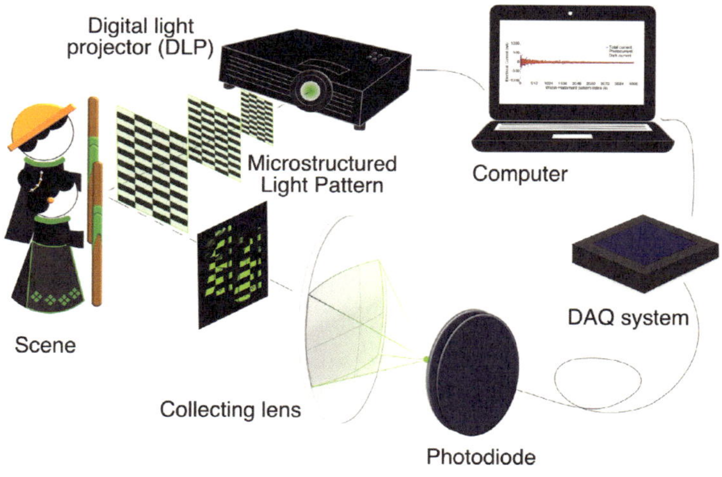

Figure 1. Schematic representation of a single-pixel camera.

modulation is required. A DMD is a microelectromechanical system device that contains a pixelated display composed by millions of tiny switchable mirrors. Each mirror is able to switch to either $\pm 12^{\circ}$ with respect to the surface normal, which corresponds to *on* or *off* states, respectively [15]. In the last years, technology advances of the DMDs have improved the performance of SPCs, for instance reducing the acquisition time. For example, the most recently DMD Discovery series developed by Texas Instruments (DLP Discovery™ 4100 Development Kit) has a resolution up to full HD (1920 × 1080 pixels) and pattern rates up to 32,500 Hz [16], which allows to do SPI at video rates.

As a result, the SPCs have been successfully applied in many different imaging areas during the last decade. Among them, we can mention infrared imaging [17, 18], terahertz imaging [19], ultrasonic imaging [20], 3D computational imaging [21, 22], imaging through scattering media [23–25], 3D and photon counting light detection and ranging (LIDAR) imaging systems [3, 26, 27], stereoscopic imaging [28], microscopy [29, 30], holography [31, 32], and ophthalmoscope imaging [33].

The set of microstructured light patterns is also an important element of the illumination stage in SPI. The light patterns commonly used for illumination are speckle patterns, binary random distributions, and functions of different basis such as noiselets [34], wavelets [35], Fourier, and Walsh-Hadamard (WH) functions. In fact, the measurement time and the resolution of the image are directly related with the properties of the light patterns. Particularly, the WH functions are easily coded on the DMD display due to the binary modulation nature of the DMD.

3. The photodiode signal in a single-pixel camera

In SPCs, the image is computationally reconstructed from the electric signal provided by the photodensor; in our case, a photodiode. In this context, the photodiode signal represents the inner product of the set of microstructured light patterns with the scene. Therefore, to analyze the quality of the image, it is convenient to study the properties of the electrical signal provided by the photodiode and its noise sources.

By definition, a photodiode is a semiconductor device that converts the optical signal into a current signal by electronic processes [36]. The electrical current of the photodiode is composed by two terms, the photocurrent (I_P) and the dark current (I_D). The first one is due to the photoelectric effect on the photodiode surface and it is given by [37],

$$I_P = R_\lambda \cdot P, \tag{1}$$

where P is the optical power level of the light source and R_λ is the photodiode responsivity $[\frac{A}{W}]$. The second current is always present in the photodiode, even without illumination, and it is originated by the thermal generation of electron-hole pairs in the Si p-n and InGaAs p-i-n photodiode layers. Four sources contribute to the dark current: the generation-recombination current in the depletion region, the diffusion current from the undepleted regions, the tunneling current, and the surface leakage current [38–40]. Nevertheless, the current-voltage of a p-n diode can be ideally described by the Shockley equation, which is given by [36].

$$I_D = I_s(T)\left[e^{\left(\frac{qV_A}{k_BT}\right)} - 1\right], \tag{2}$$

where $q = 1.602 \times 10^{-19}\ C$ is the electron charge, V_A is the bias voltage, $k_B = 1.381 \times 10^{-23}\ \frac{J}{K}$ is the Boltzmann's constant, and T is the absolute temperature. Then, under the reverse-bias condition, the saturation current (as a function of the temperature T) can be written as

$$I_s(T) = C_n A_P T^3 e^{\left(-\frac{E_g(T)}{k_BT}\right)} + C_n A_P T^{\frac{3}{2}} e^{\left(-\frac{E_g(T)}{2k_BT}\right)}. \tag{3}$$

Assuming that the temperature is lower or close to room temperature, the first term in Eq. (3) can be considered negligible [36]. Taking this into account and substituting this expression in Eq. (2), the dark current is given by

$$I_D = C_n A_P T^{\frac{3}{2}} e^{\left(-\frac{E_g(T)}{2k_BT}\right)}\left[e^{\left(\frac{qV_A}{k_BT}\right)} - 1\right], \tag{4}$$

where C_n is a constant factor $[\frac{nA}{cm^2}]$, A_P is the photodiode area $[cm^2]$, and $E_g(T)$ is the band gap energy of the photodiode material $[eV]$ as a function of the temperature. $E(T)$ is described by the Varshni empirical relation for a Si p-n photodiode case [41]

$$E_g(T) = E_g^{Si}(0) - \frac{\alpha^{Si}T^2}{T + \beta^{Si}}, \tag{5}$$

and by the Sajal Paul relation for the $In_{1-x}Ga_xAs$ p-i-n photodiode case [42]

$$E_g(x,T) = E_g^{InAs}(0) - \frac{\alpha^{InAs}T^2}{T+\beta^{InAs}} + \left[E_g^{GaAs}(0) - \frac{\alpha^{GaAs}T^2}{T+\beta^{GaAs}} - E_g^{InAs}(0) + \frac{\alpha^{InAs}T^2}{T+\beta^{InAs}}\right]x - 0.475x(1-x), \tag{6}$$

where $E_g(0)$, α and β are material constants. **Table 1** shows typical values for them [43, 44]. Finally, when the photodiode is illuminated, the total current (I) at room temperature is given by

$$\begin{aligned} I &= I_P + I_D \\ &= R_\lambda \cdot P + C_n A_P T^{\frac{3}{2}} e^{\left(-\frac{E_g(T)}{2k_BT}\right)} \left[e^{\left(\frac{qV_A}{k_BT}\right)} - 1\right]. \end{aligned} \tag{7}$$

In the single-pixel camera process, the photocurrent and the dark current signals have an associated error, due to the discrete nature of the electrical charge [45]. The noise of the former one is known as the photocurrent shot noise (σ_P) and it is given by

$$\sigma_P = \sqrt{2q\bar{I}_P B}, \tag{8}$$

where B is the noise bandwidth and $\bar{I}_P$ is the photocurrent mean value. The noise of the second current is called the dark-current shot noise (σ_D), defined as

$$\sigma_D = \sqrt{2q\bar{I}_D B}, \tag{9}$$

where $\bar{I}_D$ is the dark current mean value. Note that the photocurrent shot noise depends on the optical signal level and the dark-current shot noise does not. The sum of both noise values is known as shot noise (σ_{shot}) [37] and it follows the Poisson distribution statistics (commonly approximated by a Gaussian distribution when the current is large).

For the sake of completeness, we will consider the Johnson-Nyquist (or thermal) noise $(\sigma_{thermal})$, which is produced by the random thermal motion of electrons in a resistor, and it can be modeled as a stationary Gaussian random process (nearly white noise) [37]. The thermal noise is given by [36].

$$\sigma_{thermal} = \sqrt{\frac{4k_BTB}{R_{SH}}}, \tag{10}$$

	$E_g[eV]$ **at** $T = 0K$	$\alpha\left[\frac{eV}{K}\right]$	$\beta[K]$
Si	1.1557	7.021×10^{-4}	1108
GaAs	1.519	5.405×10^{-4}	204
InAs	0.417	2.76×10^{-4}	93

Table 1. Values of material parameters E_g, α, and β

where R_{SH} is the shunt resistance. Since σ_P, σ_D, and $\sigma_{thermal}$ are linearly independent noise sources, the total noise (σ_T) can be written as

$$\sigma_T = \sqrt{\sigma_P^2 + \sigma_D^2 + \sigma_{thermal}^2}. \tag{11}$$

Consequently, the signal-to-noise ratio (SNR) of the electrical current signal in units of decibels is defined as [37].

$$SNR = 10\log\left(\frac{\bar{I}_P^2}{\sigma_T^2}\right) = 20\log\left(\frac{\bar{I}_P}{\sqrt{2q\bar{I}_P B + 2q\bar{I}_D B + \frac{4k_B TB}{R_{SH}}}}\right). \tag{12}$$

4. Numerical model of the single-pixel camera

In this section, a numerical model of the SPC is described. The camera model takes into account the optical power level (P_{inp}) and the wavelength (λ_s) of the incident light, which is assumed to be a monochromatic light source. It also considers the photocurrent, the dark current, the photocurrent shot noise, the dark-current shot noise, and the thermal noise as a function of the photodiode parameters.

In this model, the microstructured light patterns are 2-D functions $H_k(n,n)$ pertaining to the orthonormal Walsh-Hadamard (WH) basis [46, 47]. The functions $H_k(n,n)$ are square binary matrices whose elements are equal to ± 1, where $k = 1, \ldots, N^2$ denotes the pattern index, (n,n) is the pattern spatial location, and (N,N) are the pattern spatial dimensions. For experimental reasons, in our model, $H_k(n,n)$ is considered to be composed of a positive $H_k^+(n,n)$ and a complementary part $H_k^-(n,n)$, fulfilling the following relation,

$$H_k(n,n) = H_k^+(n,n) - H_k^-(n,n). \tag{13}$$

In absence of noise, the mathematical properties of $H_k(n,n)$ allow us to recover an exact replica of the object with a 2-D spatial resolution equal to (N,N) pixels.

The numerical process developed to simulate the SPC, from the moment in which the light source illuminates the DMD up to the image reconstruction (see **Figure 1**), is as follows:

Step 1: Obtain the number of photons per second $\left(\gamma_{inp}^{k\pm}\right)$ corresponding to the wavelength (λ_s) and to the optical power of the light source (P_{inp}), arriving at the DMD,

$$\gamma_{inp}^{k\pm} = floor\left(\frac{P_{inp}}{E_\gamma}\right), \tag{14}$$

where $E_\gamma = \frac{hc}{\lambda_s}$ is the photon energy.

Step 2: Distribute the $\gamma_{inp}^{k^{\pm}}$ photons spatially following a statistical Poisson distribution in a matrix $A(n,n)$,

$$B^{k^{\pm}}(n,n) = \gamma_{inp}^{k^{\pm}} \cdot A(n,n). \tag{15}$$

Step 3: Multiply the photon matrix $B^{k^{\pm}}(n,n)$ by $H_k^+(n,n)$ and $H_k^-(n,n)$, respectively,

$$C^{k^{\pm}}(n,n) = B^{k^{\pm}}(n,n) \cdot H_k^{\pm}(n,n). \tag{16}$$

Step 4: Calculate the number of photons per second $\left(\gamma_{out}^{k^{\pm}}\right)$ that strike on the photodiode by projecting $C^{k^{\pm}}(n,n)$ onto the object $O(n,m)$ as a dot product,

$$\gamma_{out}^{k^{\pm}} = \sum_{m=1}^{M}\sum_{n=1}^{N} C^{k^{\pm}}(n,n) \cdot O(n,m). \tag{17}$$

Step 5: Obtain the optical power level ($P_{out}^{k^{\pm}}$) at the photodiode,

$$P_{out}^{k^{\pm}} = \gamma_{out}^{k^{\pm}} \cdot E_{\gamma}. \tag{18}$$

Step 6: Calculate the total current as the sum of the photocurrent and the dark current according to Eq. (7),

$$I^{k^{\pm}} = R_{\lambda} \cdot P_{out}^{k^{\pm}} + C_n A_P T^{\frac{3}{2}} e^{\left(-\frac{E_g(T)}{2k_B T}\right)} \left[e^{\left(\frac{qV_A}{k_B T}\right)} - 1\right]. \tag{19}$$

Step 7: Obtain the noisy current $\left(I_{sp}^{k\pm}\right)$ of the photodiode by adding the photocurrent shot noise, the dark-current shot noise and the thermal noise to $I^{k^{\pm}}$. The noise terms are generated following a Gaussian distribution, taking into account Eqs. (8)–(10), respectively.

Step 8: Obtain the normalized photodetector signal related to $H_k(n,n)$ (taking into account Eq. (13)) as,

$$I_{sp}^{k} = \frac{1}{I_0}\left(I_{sp}^{k^+} - I_{sp}^{k^-}\right). \tag{20}$$

where I_0 is the signal measured by the photodiode when the object is illuminated with an uniform pattern.

Step 9: Calculate the image $I(n,n)$ by multiplying the noise current signal of the photodiode by the WH functions as follows

$$I(n,n) = \frac{1}{N^2}\sum_{k=1}^{N^2} I_{sp}^{k} \cdot H_k(n,n). \tag{21}$$

5. Results

5.1. Numerical results

The numerical model described in the previous section was used to analyze the performance of a SPC formed by photodiode detectors under different circumstances. Three different studies were developed analyzing the image quality when: (1) the optical power level of the light source changes; (2) we use light sources with different wavelengths; and (3) the photodiode temperature varies. The simulations were performed for two commercial photodiodes, DET36A Thorlabs and DET10C Thorlabs, whose specifications are shown in **Table 2**. Moreover, the dark current (I_D), the dark-current shot noise (σ_D), and the thermal noise ($\sigma_{thermal}$) curves are plotted as a function of the temperature (see **Figure 2**). The curves were obtained

Parameter	Symbol	Silicon biased detector	$In_{0.53}Ga_{0.47}As$ biased detector
Detector name		DET36A Thorlabs	DET10C Thorlabs
Photodiode active area	A_P	13.0 mm^2	0.8 mm^2
Wavelength range		350–1100 nm	900–1700 nm
Band gap energy at 298 K	E_g	1.1114 eV	0.7379 eV
Rise time response	t_r	14.0 ns	10.0 ns
Noise bandwidth	B	0.025 nHz	0.035 nHz
Bias voltage	V_A	10.0 V	5.0 V
Saturation current at 298 K	I_s	0.35 nA	1.0 nA
Shunt resistance	R_{sh}	1.0 GΩ	10.0 GΩ
NEP at λ_P		$1.6 \times 10^{-14} \frac{W}{\sqrt{Hz}}$	$2.5 \times 10^{-14} \frac{W}{\sqrt{Hz}}$

Table 2. Photodiode parameters data.

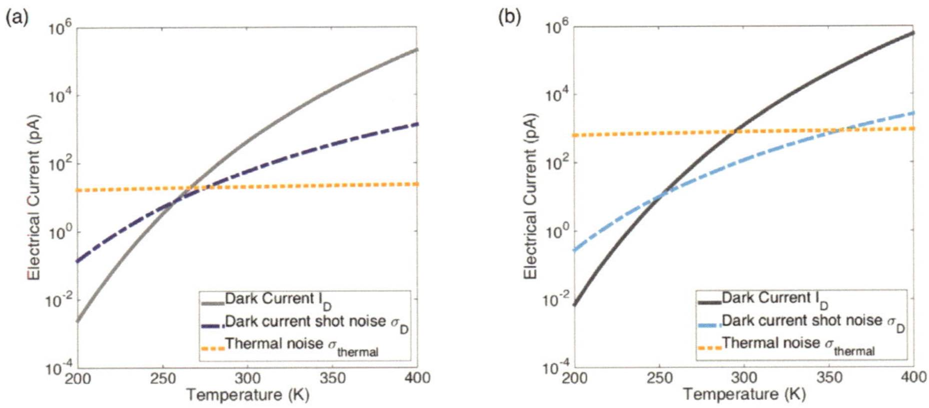

Figure 2. Simulated dark current, dark-current shot noise and thermal noise as a function of the temperature for (a) Si biased detector (DET36A Thorlabs) and (b) InGaAs biased detector (DET10C Thorlabs).

taking into account the Varshni empirical relation (Eq. (5)) for the Si detector case and the Sajal Paul relation (Eq. (6)) for the InGaAs detector case. The material parameters are shown in **Table 1**.

Firstly, we analyzed the image quality as a function of the optical power level of the light source. We fixed the wavelength of the light source to 520 nm and the photodiode temperature to 298 K. **Figure 3** shows the photocurrent, the dark current, and the total current associated to the single-pixel camera for two different optical power levels; (a) 42.29 μW and (b) 0.0085 μW,

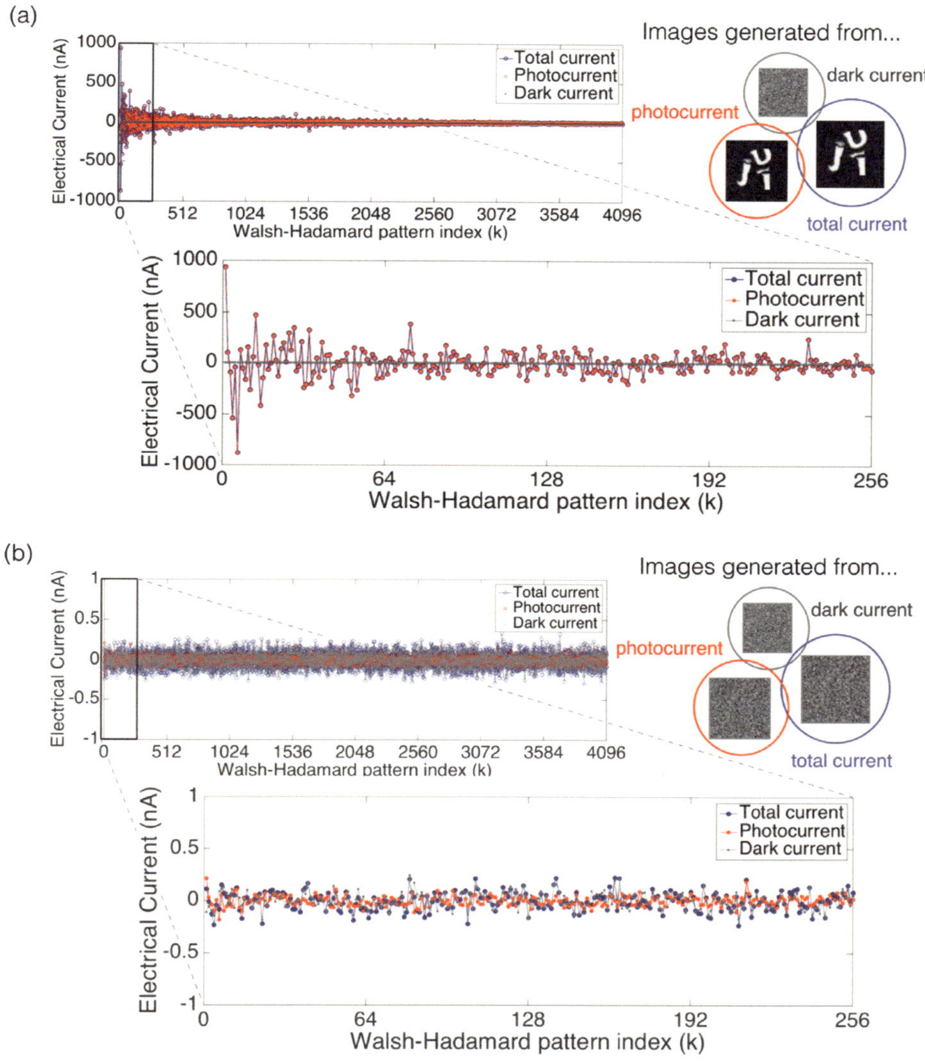

Figure 3. Photocurrent, dark current, and total current with their associated noise values as a function of the WH pattern index. Two different optical power values were considered (a) 42.49 μW and (b) 0.0085 μW. The wavelength of the light source was fixed at 520 nm and the photodiode temperature at 298 K. Images computed from these electric signals are shown on the right. In both cases, the resolution of the WH patterns is 64 $\times$ 64 pixels (Reprinted from [10]).

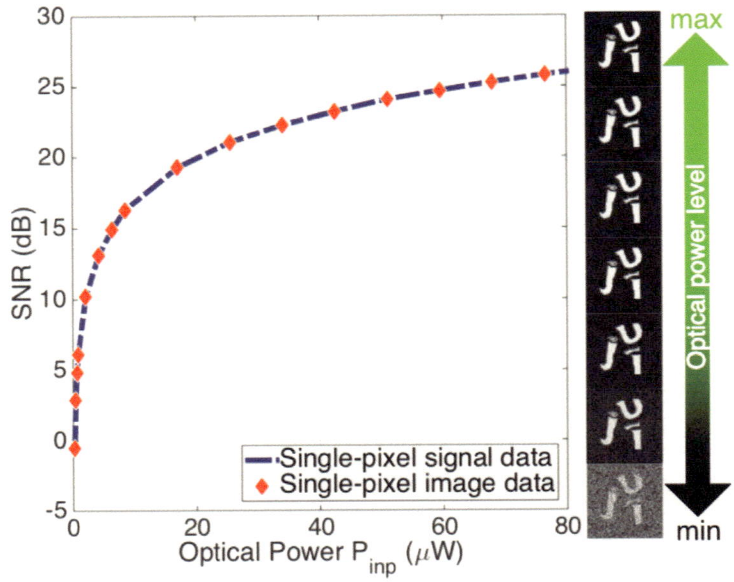

Figure 4. SNR of the signal and the recovered images as a function of the optical power P_{inp}. The recovered images on the right part are a sample of the SNR red data points (reprinted from [10]).

respectively. We also show recovered images by applying SPI techniques to the different current signals in the plot. We can see that for the case of low light power in **Figure 3(b)**, the photocurrent is noisier, the total current is worst, and therefore, the quality of the image deteriorates with respect to the case in **Figure 3(a)**.

On the other hand, we numerically evaluated the image quality using the SNR metric defined by [48]. The reference image is obtained by SPI techniques but using only the photocurrent values without considering the noise sources. Afterward, we plot in **Figure 4** the SNR as a function of the optical power of the light source. As expected, the image quality obtained by the SPC improves when the optical power increases. In the same figure, we also plot the curve of the SNR of the photodiode signal as a function of the optical power level. The reference signal is again the photocurrent signal without noise values. Of course, the SNR is the same in both cases. Therefore, we will use the SNR applied to the images from now onwards.

Second, we analyzed how the wavelength of the light source influences the performance of the SPC. The optical power of the light source was set to 8.49 μW and the photodiode temperature to 298 K. **Figure 5(a)** shows the dependence of the SNR versus wavelength for the DET36A and the DET10C photodiodes. In **Figure 5(c)**, we display several images reconstructed with our model for different wavelengths of the incident light. The key point to understand the relationship between the image quality and the wavelength is the photodiode responsivity (R_λ). In a photodetector, the incident optical power and the generated photocurrent are proportionally related by the responsivity (Eq. (1)). Therefore, the photocurrent increases as the responsivity rises up, although optical power remains constant. The responsivity versus wavelength curves

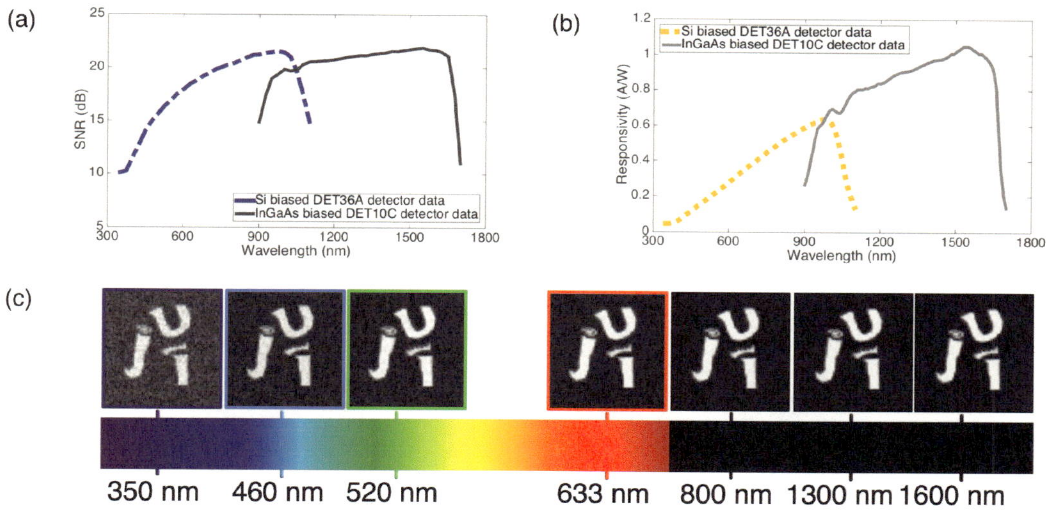

Figure 5. (a) SNR as a function of the wavelength of the light source; (b) responsivity data of both photodiodes [49, 50]; and (c) recovered images for different wavelengths.

for both photodiodes are presented in **Figure 5(b)** [49, 50]. As we can see comparing **Figure 5 (a)** and **(b)**, the behavior of the SNR and responsivity curves are closely related.

Finally, we analyzed the dependence of the image quality with the photodiode temperature. The wavelength of the light source was set to 520 and 1600 nm for the DET35A and the DET10C detectors, respectively. For each detector, three curves of the SNR as a function of the photodiode temperature are plotted for constant values of the optical power (42.49 μW, 8.49 μW and 0.21 μW) as shown in **Figure 6**. Moreover, several images for different values of temperature and optical power are displayed. In general, the SNR of the image decreases as the temperature increases. However, as we can see in the figure, the influence of the temperature on the SNR value is less significant for higher optical power levels. In particular, the performance of these photodiodes is suitable even with high temperatures whenever the optical power is higher than 8.50 μW. As previously shown in **Figure 2**, the dark current and the dark-current shot noise increase when the temperature increases. Although the current and its noise increase when the temperature increases, this effect is negligible in the SNR curves if the optical power level is high.

5.2. Experimental results

A scheme of the experimental setup by using transillumination is depicted in **Figure 7**. In this case, a DMD (DLP Discovery 4100, Texas Instrument) was illuminated with a collimated light beam generated with an incoherent white-light source. A narrow band pass filter (P10-515-S 93819, Corion) centered at a wavelength of 520 nm with a bandwidth of 20 nm was used to avoid spectral artifacts. In order to apply SPI techniques, microstructured light patterns corresponding to 2-D functions of the orthonormal WH basis with 64 × 64 pixels were coded on the DMD display in a chip area of 1024 × 1024 micromirrors with a micromirror pixel pitch

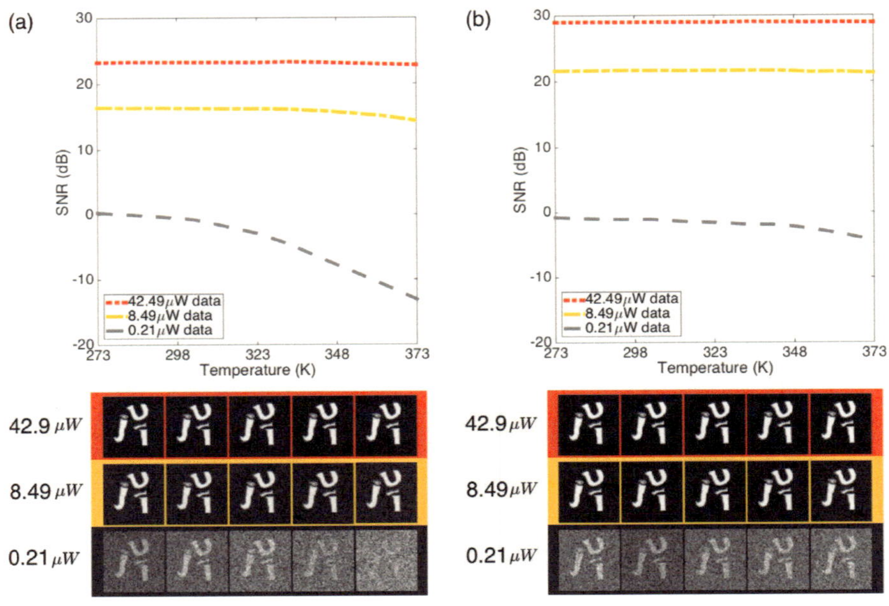

Figure 6. SNR dependence with the photodiode temperature for three optical power levels: 42.49 μW, 8.49 μW, and 0.21 μW for (a) the Si biased detector (DET36A Thorlabs), and (b) the InGaAs biased detector (DET10C Thorlabs). The recovered images obtained for those optical power levels are shown as well. For those images, the temperature range starts at 273 K and ends at 373 K in 25 K steps.

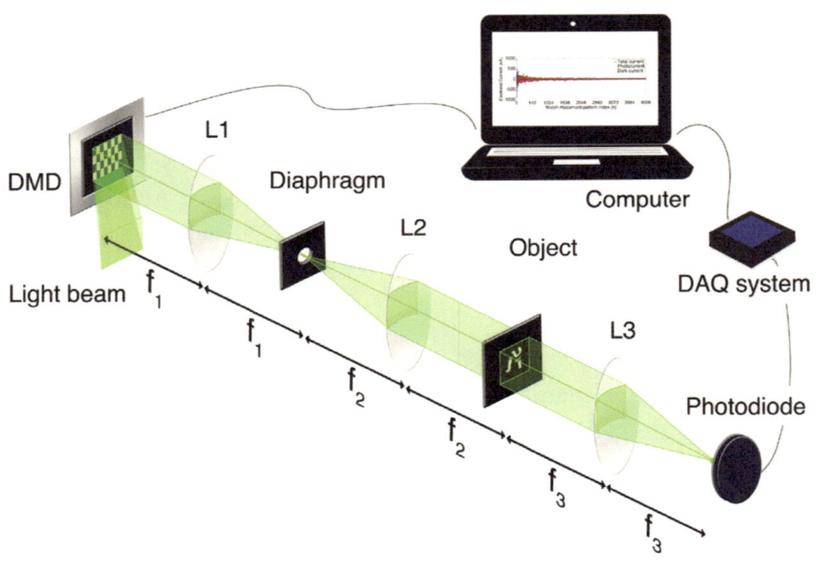

Figure 7. Experimental setup of the single-pixel camera.

of 10.8 μm. The WH patterns were projected onto the object plane using a 4-f optical imaging system formed by two achromatic lenses L1 and L2. The focal distances of L1, and L2 were $f_1 = 100$ mm, and $f_2 = 100$ mm, respectively. The magnification factor of the 4-f optical system was 1.0; therefore, the field of view (FOV) was 1.10 × 1.10 cm, which is, in fact, the size of the WH patterns on the DMD display. Note that a circular diaphragm was used on the Fourier plane in order to filter unwanted diffracted orders produced by the periodic micromirror arrangement on the DMD display. The light transmitted by the object was subsequently collected by a lens L3 and focused onto a Si biased detector (DET36A Thorlabs) located at the back focal point of L3; $f_3 = 50$ mm. The optical power level of the incident light was adjusted by using a neutral density filter wheel (NDC-100S-4M-Mounted Step Variable ND Filter) located in front of the lamp and measuring power with a power meter (Coherent, FieldMaster GS) close to the photodiode sensor. Finally, the signal was digitized and saved in a computer by using a DAQ system. The image was reconstructed by using Eq. (21). The object was a black and white logo of our university (UJI) printed in a transparent acetate slide. The object size was 1.10 × 1.10 cm with a total transmittance factor of 0.12. It should be mentioned that this object has the same features as the one used in Sub Section 5.1.

Figure 8(a) shows numerically and experimentally recovered images with different levels of the optical power. We can see that, in both cases, the noise level decreases with the optical power. This is corroborated by the results in **Figure 8(b)**, which show that the SNR curve corresponding to images obtained with the simulated and the experimental systems have a similar dependence with the optical power. This fact confirms the validity of our numerical model.

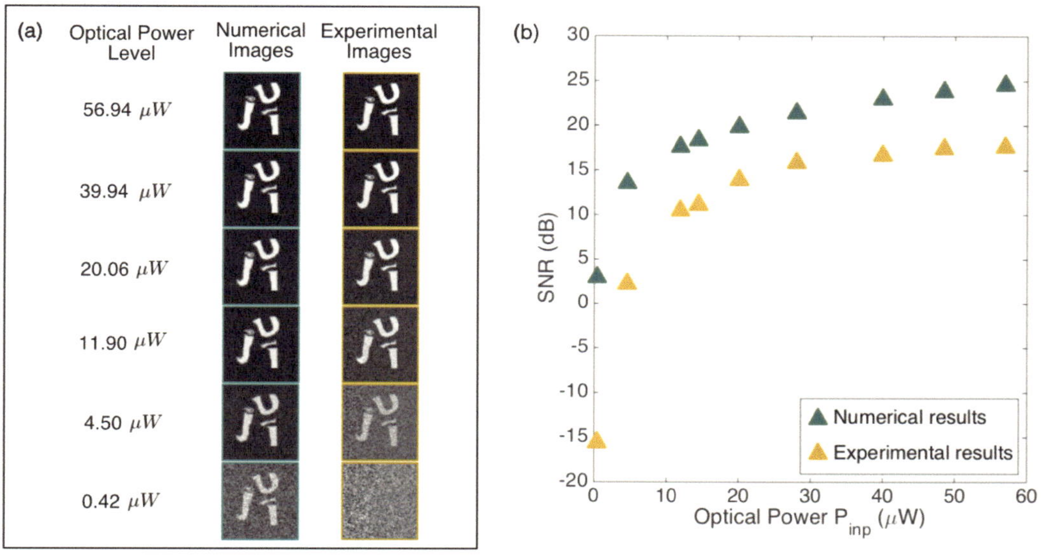

Figure 8. (a) Numerically and experimentally recovered images for different optical power levels P_{inp}; (b) SNR dependence with P_{inp} for the experimental and numerical images.

However, even though the model has been developed taking into account the most important noise factors during the imaging process, there is still a discrepancy in the values of the SNR for the experimental and the simulated images. This difference is produced by several other noise sources not included in the model. First, we have considered that both the DMD reflectance and the object transmittance are ideal binary functions, which is not true in practice. Second, we did not introduce background, or ambient, light into the numerical model, with the unavoidable associated noise. Finally, we did not consider the current-to-voltage and the analog-to-digital conversion processes, which produce certain amount of noise. A clear example in the last case is the quantization noise.

6. Conclusions

In this chapter, a numerical model of a single-pixel camera (SPC) has been developed, considering the characteristics of the incident light and the physical properties, as well as the specifications of the photodiode. We have accomplished a careful and rigorous mathematical review of the electrical behavior of Si and InGaAs detectors. Our model takes into account the photocurrent, the dark current, the photocurrent shot noise, the dark-current shot noise, and the Johnson-Nyquist (thermal) noise of two commercial photodiodes, a Si and an InGaAs photodetector.

Numerical simulations with our model have allowed us to analyze the behavior of the single-pixel imaging (SPI) technique in different contexts. In particular, we have studied the quality of the final image as a function of the power of the light source. We have corroborated the reduction of the SNR for low light levels. We have also observed the clear link between the quality of the photocurrent signal and the quality of the reconstructed image. These results can be useful to predict the behavior of imaging systems working in low light level conditions.

We have also studied the dependence of the SNR with the wavelength of the light source. In this case, we conclude that the influence of the wavelength arises from the variation of the quantum efficiency and the responsivity of the photodetector. Such analysis could be the first step in the application of SPI techniques to multispectral imaging.

Finally, we have analyzed the quality of the images provided by the SPC as a function of the photodiode temperature. The study is performed for both a Si biased and an InGaAs biased detector. The main conclusion in this case is that the SNR of the reconstructed images changes only slightly with the temperature for high values of the light power. However, the reduction is clearly significant for low light levels. Therefore, cooling the detector can play an important role in photon counting or low light level applications.

An experimental SPC has been developed in order to validate the results provided by our model. The quality of the images obtained experimentally does not match perfectly with that predicted by the model. The discrepancy is due to several unaccounted sources of uncertainty such as nonuniformities in the mirrors of the DMD or in the object substrate, as well as noise introduced in the signal amplification process or the analog to digital conversion procedure.

However, we have shown that the quality of the final image, in terms of the SNR, changes in a similar way with the light power. This allows us to confirm that the model can be useful to predict the behavior of SPI systems based on photodiodes under different circumstances.

Funding information

We acknowledge financial support from MINECO (FIS2016-75618-R and FIS2015-72872-EXP), Generalitat Valenciana (PROMETEO/2016/079), and Universitat Jaume I (P1-1B2015-35). Yessenia Jauregui-Sánchez acknowledges the Santiago Grisolía support from Generalitat Valenciana (GRISOLIA/2015/037).

Author details

Yessenia Jauregui-Sánchez[1]*, Pere Clemente[1,2], Pedro Latorre-Carmona[3], Jesús Lancis[1] and Enrique Tajahuerce[1]

*Address all correspondence to: jauregui@uji.es

1 GROC·UJI, Institute of New Imaging Technologies (INIT), Universitat Jaume I, Castelló, Spain

2 Servei Central d'Instrumentació Científica (SCIC), Universitat Jaume I, Castelló, Spain

3 eVIS, Institute of New Imaging Technologies (INIT), Universitat Jaume I, Castelló, Spain

References

[1] Duarte MF, Davenport MA, Takbar D, Laska JN, Sun T, Kelly KF, Baraniuk RG. Single-pixel imaging via compressive sampling. IEEE Signal Processing Magazine. 2008;**25**:83-91

[2] Studer V, Bobin J, Chahid M, Mousavi HS, Candes E, Dahan M. Compressive fluorescence microscopy for biological and hyperspectral imaging. Proceedings of the National Academy of Sciences of the United States of America. 2012;**109**:E1679-E1687

[3] Howland GA, Lum DJ, Ware MR, Howell JC. Photon counting compressive depth mapping. Optics Express. 2013;**21**:23822-23837

[4] Soldevila F, Irles E, Durán V, Clemente P, Fernández-Alonso M, Tajahuerce E, Lancis J. Single-pixel polarimetric imaging spectrometer by compressive sensing. Applied Physics B: Lasers and Optics. 2013;**113**:551-559

[5] Durán V, Clemente P, Fernández-Alonso M, Tajahuerce E, Lancis J. Single-pixel polarimetric imaging. Optics Letters. 2012;**37**:824-826

[6] Rodríguez AD, Clemente P, Irles E, Tajahuerce E, Lancis J. Resolution analysis in computational imaging with patterned illumination and bucket detection. Optics Letters. 2014; **39**:3888-3891

[7] Xu YK, Liu WT, Zhang EF, Li Q, Dai HY, Chen PX. Is ghost imaging intrinsically more powerful against scattering? Optics Express. 2015;**23**:32993-33000

[8] Candès EJ, Walkin MB. An introduction to compressive sampling. IEEE Signal Processing Magazine. 2008;**25**:21-30

[9] Willett RM, Marcia RF, Nichols JM. Compressed sensing for practical optical imaging systems: A tutorial. Optical Engineering. 2011;**50**:072601

[10] Jauregui-Sánchez Y, Clemente P, Latorre-Carmona P, Tajahuerce E, Lancis J. Signal-to-noise ratio of single-pixel cameras based on photodiodes. Applied Optics. 2018;**57**:B67-B73

[11] Golay MJE. Multi-Slit Spectrometry. Journal of the Optical Society of America. 1949;**39**: 437-444

[12] Decker JA. Hadamard–transform image scanning. Applied Optics. 1970;**9**:1392-1395

[13] Wakin MB, Laska JN, Duarte MF, Baron D, Sarvotham S, Takhar D, Kelly KF, Baraniuk RG. An architecture for compressive imaging. In: 2006 International Conference on Image Processing IEEE, 1273-1276; 2006

[14] Clemente P, Sistemes formadors dÍmatges multidimensionals amb detecció integrada [thesis]. Spain: Universitat Jaume I; 2015

[15] Ren YX, Lu RD, Gong L. Tailoring light with a digital micromirror device. Annalen der Physik. 2015;**527**:447-470

[16] DLP Discovery 4100 Development Kit, Texas Instruments. 2018 [Online]. Available from: http://www.ti.com/tool/DLPD4X00KIT# [Accessed: June 11, 2018]

[17] Chen H, Xi N, Song B, Chen L, Zhao J, Lai KWC, Yang R. Infrared camera using a single nano-photodetector. IEEE Sensors Journal. 2013;**13**:949-958

[18] Gibson GM, Sun B, Edgar MP, Phillips DB, Hempler N, Maker GT, Malcolm GPA, Padgett MJ. Real-time imaging of methane gas leaks using a single-pixel camera. Optics Express. 2017;**25**:2998-3005

[19] Stantchev RI, Sun B, Hornett SM, Hobson PA, Gibson GM, Padgett MJ, Hendry E. Noninvasive, near-field terahertz imaging of hidden objects using a single-pixel detector. Science Advances. 2016;**2**:e1600190

[20] Huynh N, Zhang E, Betcke M, Arridge S, Beard P, Cox B. Single-pixel optical camera for video rate ultrasonic imaging. Optica. 2016;**3**:26-29

[21] Sun B, Edgar MP, Bowman R, Vittert LE, Welsh S, Bowman A, Padgett MJ. 3D computational imaging with single-pixel detectors. Science. 2013;**340**:844-847

[22] Hunt J, Driscoll T, Mrozack A, Lipworth G, Reynolds M, Brady D, Smith DR. Metamaterial apertures for computational imaging. Science. 2013;**339**:310-313

[23] Tajahuerce E, Durán V, Clemente P, Irles E, Soldevila F, Andrés P, Lancis J. Image transmission through dynamic scattering media by single-pixel photodetection. Optics Express. 2014;**22**:16945-16955

[24] Durán V, Soldevila F, Irles E, Clemente P, Tajahuerce E, Andrés P, Lancis J. Compressive imaging in scattering media. Optics Express. 2015;**23**:14424-14433

[25] Shin J, Bosworth BT, Foster MA. Single-pixel imaging using compressed sensing and wavelength-dependent scattering. Optics Letters. 2016;**41**:886-889

[26] Kirmani A, Venkatraman D, Shin D, Colaço A, Wong FNC, Shapiro JH, Goyal VK. First-Photon Imaging. Science. 2014;**343**:58-61

[27] Sun MJ, Edgar MP, Gibson GM, Sun B, Radwell N, Lamb R, Padgett MJ. Single-pixel three-dimensional imaging with time-based depth resolution. Nature Communications. 2016;**7**:12010

[28] Salvador-Balaguer E, Clemente P, Tajahuerce E, Pla F, Lancis J. Full-color stereoscopic imaging with a single-pixel photodetector. Journal of Display Technology. 2016;**12**:417-422

[29] Radwell N, Mitchell KJ, Gibson GM, Edgar MP, Bowman R, Padgett MJ. Single-pixel infrared and visible microscope. Optica. 2014;**1**:285-289

[30] Rodríguez AD, Clemente P, Tajahuerce E, Lancis J. Dual-mode optical microscope based on single-pixel imaging. Optics and Lasers in Engineering. 2016;**82**:87-94

[31] Brady DJ, Choi K, Marks DL, Horisaki R, Lim S. Compressive Holography. Optics Express. 2009;**17**:13040-13049

[32] Martínez-León L, Clemente P, Mori Y, Climent V, Lancis J, Tajahuerce E. Single-pixel digital holography with phase-encoded illumination. Optics Express. 2017;**25**:4975-4984

[33] Lochocki B, Gambín A, Manzanera S, Irles E, Tajahuerce E, Lancis J, Artal P. Single pixel camera ophthalmoscope. Optica. 2016;**3**:1056-1059

[34] Pastuszczak A, Szczygieł B, Mikołajczyk M, Kotyński R. Efficient adaptation of complex-valued noiselet sensing matrices for compressed single-pixel imaging. Applied Optics. 2016;**55**:5141-5148

[35] Rousset F, Ducros N, Farina A, Valentini G, Andrea CD′, Peyrin F. Adaptive basis scan by wavelet prediction for single-pixel imaging. IEEE Transactions on Computational Imaging. 2018;**4**:284-294

[36] Sze SM, Ng KK. Physics of Semiconductor Devices. 3rd ed. Hoboken, New Jersey: John Wiley & Sons Inc; 2007. ISBN-13: 978-0471143239

[37] Agrawal GP. Fiber-Optic Communication Systems. 4th ed. Rochester, New York: John Wiley & Sons Inc; 2010. ISBN-13: 978-0470505113

[38] Pearsall T, Piskorski M, Brochet A, Chevrier J. A $Ga_{0.47}In_{0.53}As$/InP heterophotodiode with reduced dark current. IEEE Journal of Quantum Electronics. 1981;**17**:255-259

[39] Ohnaka K, Kubo M, Shibata J. A Low Dark Current InGaAs/InP p-i-n Photodiode with Covered Mesa Structure. IEEE Transactions on Electron Devices. 1987;**34**:199-204

[40] Jacob B, Witzigmann B, Klemenc M, Petit C. A TCAD methodology for high-speed photodetectors. Solid-State Electronics. 2005;**49**:1002-1008

[41] Varshni YP. Temperature dependence of the energy gap in semiconductors. Physica. 1967; **34**:149-154

[42] Paul S, Roy JB, Basu PK. Empirical expressions for the alloy composition and temperature dependence of the band gap and intrinsic carrier density in $Ga_xIn_{1-x}As$. Journal of Applied Physics. 1991;**69**:827-830

[43] Konnik M, Welsh J. High-level numerical simulations of noise in CCD and CMOS photosensors: Review and tutorial . arXiv Preprint. arXiv. 2014;**1412**:4031

[44] Vurgaftman I, Meyer JR, Ram-Mohan LR. Band parameters for III-V compound semiconductors and their alloys. Journal of Applied Physics. 2001;**89**:5815-5875

[45] Bennett WR. Sources and Properties of Electrical Noise. Electrical Engineering. 1954;**73**: 1001-1006

[46] Paley REAC. On orthogonal matrices. Journal of Mathematical Physics. 1933;**12**:311-320

[47] Pratt W, Kane J, Andrews H. Hadamard transform image coding. Proceedings of the IEEE. 1969;**57**:58-68

[48] Gonzalez RC, Woods RE. Digital Image Processing. 3rd ed. New Jersey: Prentice Hall; 2008. ISBN-13: 978-0131687288

[49] Si Biased Detector DET36A Thorlabs, Inc. 2018 [Online]. Available from: https://www.thorlabs.com/thorproduct.cfm?partnumber=DET36A [Accessed: June 21, 2018]

[50] InGaAs Biased Detector DET10C Thorlabs, Inc. 2018 [Online]. Available from: https://www.thorlabs.com/thorproduct.cfm?partnumber=DET10C [Accessed: June 21, 2018]

Chapter 7

Overcoming the Bandwidth-Quantum Efficiency Trade-Off in Conventional Photodetectors

Tianyi Zhou and Kuan W.A. Chee

Additional information is available at the end of the chapter

http://dx.doi.org/10.5772/intechopen.86506

Abstract

Optical systems and microwave photonics applications rely heavily on high-performance photodetectors having a high bandwidth-efficiency product. The main types of photodetector structures include Schottky and PIN-photodiodes, heterojunction phototransistors, avalanche photodetectors, and metal-semiconductor-metal photodetectors. Vertically-illuminated photodetectors intrinsically present bandwidth-efficiency limitations, but these have been mitigated by new innovations over the years in quantum well photodetectors, edge-coupled photodetectors and resonant-cavity enhanced photodetectors for improved photophysical characteristics. Edge-coupled ultra-high-speed photodetectors have yielded high conversion efficiencies, and the active device structure of resonant-cavity-enhanced photodetectors allows wavelength selectivity and optical field enhancement due to resonance, enabling photodetectors to be made thinner and hence faster, while simultaneously increasing the quantum efficiency at the resonant wavelengths. Single-photon avalanche diodes have been developed, which combine an ultimate sensitivity with excellent timing accuracy. Further advances in addressing the bandwidth-quantum efficiency trade-off have incorporated photon-trapping nanostructures and plasmonic nanoparticles. Nanowire photodetectors have also demonstrated the highest photophysical performance to date.

Keywords: bandwidth-efficiency product, saturation current, quantum efficiency, photosensitivity, optical absorption, drift layers

1. Introduction

High-performance photodetectors (PDs) are key components in optical systems and microwave photonics applications. Examples include radio telescope arrays, optical fiber communication systems and optically controlled phased array radar. Over the past several decades, the design principles of PDs and their technologies have become well developed, as various structures and fabrication/processing strategies have been established. Overall, the main types of PDs include *p-i-n* PDs, metal-semiconductor-metal (MSM) PDs, waveguide PDs (WGPDs) and traveling-wave PDs (TWPDs). These can be placed into three categories, according to the direction of optical propagation in the PDs, i.e., vertically-illuminated PDs (VPDs), edge-coupled PDs (EC-PDs) and resonant-cavity enhanced PDs (RCE-PDs). On the other hand, the lump and distributed PDs can be classified based on the component properties. The basic requirements for the PDs are high efficiency and high bandwidth, which are especially significant for systems operating at high data rates. In general, the quality of the different types of the high-speed PDs is characterized by the bandwidth-efficiency product. Another performance requirement of PDs is a high saturation current, especially for high power systems.

2. Vertically-illuminated photodetectors (VPDs)

The VPD comprises either the *p-i-n* or MSM structure. Upon optical illumination, electron-hole pairs generated in the device are separated by the electric fields within the *i*-region, thus contributing to a photocurrent through the processes of drift and diffusion. Simple-structured *p-i-n* PDs are the most common components in many optical systems. Yet, in order to improve on existing features of the conventional *p-i-n* PDs, different design variations, such as, those found in dual-depletion-region photodiodes (DDR PDs) [1, 2], uni-traveling-carrier photodiodes (UTC-PDs) [3–5] and avalanche photodiodes (APDs) [6–9], were extensively studied. Utilizing optical absorption layers combined with drift layers having wide bandgap, the DDR PDs typically have a larger bandwidth-efficiency product than that of conventional *p-i-n* PDs. In addition, the saturation current can be increased by optimizing the thicknesses of the absorption and drift layers [10]. To increase both bandwidth and saturation current, the UTC *p-i-n* structure is used, via leveraging the fast electrons during charge carrier transport. Thanks to the internal gain based on the avalanche multiplication effect, an enhanced sensitivity can be achieved by the APDs at the expense of higher operating voltages. MSM PDs based on the Schottky barrier [11–13] are another type of VPDs, which possess a smaller capacitance and lower dark current compared with that of the traditional design.

Due to broad and significant military and civilian applications, research on infrared detection and infrared photodetectors has intensified. In past decades, work on developing the operating temperature and spectral sensitivity capabilities of infrared photodetectors have become significant with the rapid development of photoelectric materials, for example, mercury cadmium telluride (HgCdTe) ternary alloys. Since the first synthesis of HgCdTe materials [14], HgCdTe infrared detectors with variable wavelength response have been manufactured by varying the alloy composition [15]. The amount of cadmium in the alloy can be selected in order to tune the bandgap which in turn determines the optical absorption of the material in the desired infrared range spanning the shortwave infrared to the very long wave infrared. As reported in [16, 17], HgCdTe infrared detectors with low frequency noise and high R_0A product in the long wavelength spectral region were demonstrated at liquid nitrogen temperatures. As a result of large optical coefficients, more than 70% quantum efficiency has been achieved in HgCdTe infrared photodetectors [18].

Although HgCdTe is considered as an ideal material providing high degrees of freedom in infrared detector design, the difficulty in the fabrication and integration of such narrow bandgap materials (0–1.5 eV) is one practical limitation toward developing large-scale array applications [15]. Alternatively, photodetectors employing quantum wells in wide bandgap semiconductors (e.g., III-nitrides) were studied, such as, the so-called quantum well infrared photodetectors (QWIPs). Taking advantage of the artificial quantum well structure, the photocurrent is derived from optical absorption due to intersubband transitions involving many interacting and quantum-confined electrons. Based on previous theoretical and experimental investigations [19–22], Levine et al. [23] demonstrated the first QWIP, achieving a high peak responsivity at a wavelength of 10.8 μm. Thereafter, QWIPs were extensively explored [24–28] and related applications were developed [29–31].

However, n-type doped QWIP cannot utilize normal incidence illumination, and therefore optical coupling can be realized using gratings [32, 33], corrugated surfaces or 45° edge illumination [34, 35] to achieve promising results. Despite the relatively low quantum efficiency, the high uniformity and excellent reproducibility benefitting from mature growth and processing technologies represent main advantages of the QWIP over previous generation infrared detectors. It is the superior QWIP technology that makes large-scale focal plane arrays (FPA) possible. Examples include 1024 × 1024 pixel QWIP FPAs at mid-wavelength infrared and long-wavelength infrared [29], and 640 × 512 pixel four-band FPAs fabricated by monolithic stacking of different multi-quantum well structures [36, 37].

3. Edge-coupled photodetectors

Although various structures have been proposed and experimentally characterized, the bandwidth-efficiency product of conventional VPDs are limited due to the trade-off between quantum efficiency and bandwidth, which imposes a limit on the speed and sensitivity for photonic applications. For VPDs, increasing the thickness of the PD absorption layer offers the advantages of high quantum efficiency but suffers from a narrow bandwidth. Fortunately, the edge-coupled WGPD has been widely investigated as a promising approach to overcome the bandwidth-efficiency trade-off found in the VPD. The structure of the WGPD permits the bandwidth and efficiency to be specified almost independently because the quantum efficiency is determined by the waveguide length instead of the absorption layer thickness. However, the optical waveguide structure of the WGPD results in a low optical coupling efficiency [38], which is mainly caused by the mode mismatch between waveguide and optical fiber. In practice, efficient coupling is usually enhanced by a mode field converter [39]. Accordingly, depending on the structural configuration, WGPDs can be divided into mushroom-WGPDs and TWPDs.

As reported in [40], a bandwidth of 28 GHz and an efficiency of 25% have been achieved by the first ever high-speed edge-coupled WGPD. In 1991, WGPDs with double-core multimode waveguide structures were proposed to address the coupling problem [41, 42]. The calculated coupling efficiency of the WGPD having such a structure can exceed 80% [43], which is regarded as a breakthrough in WGPDs for practical applications. By combining the structures of the waveguide and photodiode, the waveguide-fed photodiode (WG-fed-PD) is another design innovation to boost the coupling efficiency of the edge-coupled WGPD. Besides, the WG-fed-PD is ideal for implementation in optoelectronic integrated circuits. Previously, 70-GHz and 100-GHz photodetectors based on WG-fed-PD have been reported in [44, 45], respectively. Since WGPDs are categorically lumped devices, their bandwidths are limited by the RC time introduced by the parasitic capacitances and resistances. Kato et al proposed a new structure, which is the so-called the mushroom-WGPD having cladding layers that are wider than the core layer [46]. In such a structure, the capacitance as well as contact resistance can be reduced to obtain a larger bandwidth. In [47], a mushroom-WGPD with a bandwidth-efficiency product of 55 GHz was demonstrated. Furthermore, the distributed-element TWPD was proposed to overcome the RC bandwidth limitation of the WGPD. Although the structures of TWPD and WGPD are similar, the electrical properties of these two photodetectors are essentially different. Therefore, the TWPD bandwidth is mainly limited by the mismatch of the optical wave and microwave propagation velocities rather than the RC time delay.

As early as 1990, the design concept of the TWPD was reported by Taylor et al. [48], and a velocity-matched *p-i-n* TWPD [49] was proposed soon after. Since the first TWPD was experimentally demonstrated in 1994 [50], TWPDs with different configurations have been extensively studied [51, 54]. The photodiode element used in the TWPD can be a *p-i-n*, MSM diode [52] or avalanche diode. The TWPD structures are configured in various forms, in which the PD is based on the simultaneous operation of optical and electrical waveguides. Additionally, the photodiode elements can be distributed over the length of the waveguides. The so-called periodic TWPD or velocity-matched distributed photodetector (VMDP) is designed based on such a structure, where the optical waveguide is periodically loaded by discrete photodiodes [51, 53].

4. Resonant-cavity-enhanced photodetectors

As stated earlier, it is possible to mitigate the limited bandwidth-efficiency product in VPDs by means of increasing the length of the optical paths while retaining the thickness of the absorption layer. Thus, the resonant-cavity-enhanced photodetector (RCE-PD) was put forth as an alternative method to solve the trade-off conundrum between efficiency and bandwidth. Since the 1990s, a family of RCE-PDs was proposed, in which the photophysical performance was enhanced by placing the VPD within a Fabry-Perot resonator [55]. Since the photodiode elements incorporated inside the resonator are conventional VPDs, it should be noted that the electrical parameters of the RCE-PD, such as, the bandwidth, and dark and saturation currents, will not be enhanced. Based on microring resonators, Abaeiani et al. presented a new structure called the RCE-WGPD or microring PD (MRPD) [56], taking advantage of both the RCE-PDs and WGPDs. With such a structure, selective wavelength detection as well as a high efficiency-bandwidth product can be achieved. Without the mirrors used in traditional

http://dx.doi.org/10.5772/intechopen.86506

RCE-PDs, the MRPDs are suitable for planar lightwave circuit integration. Various photosensitive devices based on MRPDs were reported in [57–59]. Moreover, the RCE-PDs based on grating were also presented in [60–62]. Due to the advantage of ultimate sensitivity combined with excellent timing accuracy, single-photon detectors, especially the single-photon avalanche diodes (SPADs), are important [63, 64]. As reported in [65, 66], the first RCE-SPAD was fabricated on a reflecting silicon-on-insulator (SOI) substrate.

5. Micro/nanostructured photodetectors

By adopting micro/nanostructures, photon-material interactions can be enhanced to address the trade-off between speed (bandwidth) and efficiency [67, 68]. The low-dimensional structures are able to control light for further interaction with the absorbing materials, excite the lateral propagation mode, and reduce surface reflection. Recently, silicon SPADs incorporating photon-trapping nanostructures were demonstrated [69]. Through diffraction of the vertically incident photons into the horizontal waveguide mode, the photons are trapped in the inverted pyramidal thin-film, and the absorption length is significantly increased to enhance the photon detection efficiency while retaining a low timing jitter. Similarly, a photon-trapping photodiode with micron- and nanoscale holes has demonstrated high-speed/high-efficiency performance [70], achieving an ultrafast impulse response of 30 ps FWHM (full-width at half-maximum), and a high efficiency of more than 50%. Another alternative technology being exploited to realize light-trapping in thin-film PDs is plasmonic nanostructures [71–74]. Unlike the photon-trapping mechanism enabled by micro/nanoholes, the metallic nanoparticles in plasmonic nanostructures act as sub-wavelength scattering centers, which allow coupling of the incident light into the semiconductor.

With the development of advanced nanofabrication technologies, photodetectors with integrated nanowires, i.e., nanowire PDs, have been realized and studied extensively [75–79]. In particular, several demonstrations of high-speed nanowire PDs were reported. In [80], a photoconductor with intersecting InP nanowires was demonstrated to obtain a pulse response of 14 ps FWHM at 780-nm wavelength irradiation. Compared with using bare core nanowires, higher response was achieved in MSM PDs using Schottky-contacted GaAs/AlGaAs core/shell nanowires [81]. In [82], nanopillar-based APDs have exhibited a 200-GHz gain bandwidth product at 1060-nm illumination.

6. Conclusion

This chapter introduces the main types of PD structures including the Schottky and PIN PDs, APDs, MSM PDs, and heterojunction phototransistors. Vertically-illuminated PDs have inherently low bandwidth-efficiency products but have been mitigated by new innovations in QWIP, edge-coupled, RCE and nanostructure, designs. Since the 1990s, RCE and WG PDs have been explored to address the bandwidth-quantum efficiency trade-off. RCE-SPADs have been recently developed for the ultimate in sensitivity while maintaining a low timing jitter.

CMOS- and lithography-compatible processes have been adopted in the design of SOI-based SPADs. Photons can be diffracted, guided and absorbed in different pixels, especially for tightly-patterned silicon photomultipliers. Nanostructured materials and nanoplasmonics have been exploited for enhanced photon trapping, coupling and absorption in MSM PDs and APDs, for the highest bandwidth-efficiency product.

Author details

Tianyi Zhou[1] and Kuan W.A. Chee[1,2*]

*Address all correspondence to: kuan.chee@cantab.net

1 Faculty of Electrical Engineering and Computer Science, Ningbo University, Ningbo, Zhejiang, People's Republic of China

2 Laser Research Institute, Shandong Academy of Sciences, Qingdao, Shandong, People's Republic of China

References

[1] Effenberger F, Joshi A. Ultrafast, dual-depletion region, InGaAsIlnP *p-i-n* detector. Journal of Lightwave Technology. 1996;**14**(8):1859-1864

[2] Pereira JT, Torres JPN. Frequency response optimization of dual depletion InGaAs/InP PIN photodiodes. Photonic Sensors. 2016;**6**(1):63-70

[3] Shimizu N, Watanabe N, Furuta T, Ishibashi T. InP-InGaAs uni-traveling-carrier photodiode with improved 3-dB bandwidth of over 150 GHz. IEEE Photonics Technology Letters. 1998;**10**(3):412-414

[4] Ito H, Furuta T, Kodama S, Watanabe N, Ishibashi T. InP/InGaAs uni-travelling-carrier photodiode with 220 GHz bandwidth. Electronics Letters. 1999;**35**(18):1556-1557

[5] Ishibashi T, Muramoto Y, Yoshimatsu T, Ito H. Unitraveling-carrier photodiodes for terahertz applications. IEEE Journal of Selected Topics in Quantum Electronics. 2014;**20**(6):79-88

[6] Campbell JC, Tsang WT, Qua GJ, Bowers JE. InP/InGaAsP/InGaAs avalanche photodiodes with 70-GHz gainbandwidth product. Applied Physics Letters. 1987;**51**(18):1454-1456

[7] Lahrichi M, Glastre G, Derouin E, et al. 240-GHz gain-bandwidth product back-side illuminated AlInAs avalanche photodiodes. IEEE Photonics Technology Letters. 2010;**22**(18):1373-1375

[8] Duan N, Liow TY, Lim AE, et al. 310 GHz gain-bandwidth product Ge/Si avalanche photodetector for 1550 nm light detection. Optics Express. 2012;**20**(10):11031

[9] Wu H, Wu W, Zhang H, Chen Y, Wu Z, Wang G, et al. All AlGaN epitaxial structure solar-blind avalanche photodiodes with high efficiency and high gain. Applied Physics Express. 2016;**9**(5):052103

[10] Williams KJ. Comparisons between dual-depletion-region and uni-travelling-carrier *p-i-n* photodetectors. IEE Proceedings-Optoelectronics. 2002;**149**(4):131-137

[11] Soole JB, Schumacher H. InGaAs metal-semiconductor-metal photodetectors for long wavelength optical communications. IEEE Journal of Quantum Electronics. 1991;**27**(3):737-752

[12] Colace L, Masini G, Galluzzi F, Assanto G, Capellini G, Di Gaspare L, et al. Metal-semiconductor-metal near-infrared light detector based on epitaxial Ge/Si. Applied Physics Letters. 1998;**72**(24):3175-3177

[13] Ciftcioglu B, Zhang J, Sobolewski R, Wu H. An 850-nm normal-incidence germanium metal-semiconductor-metal photodetector with 13-GHz bandwidth and 8-μA dark current. IEEE Photonics Technology Letters. 2010;**22**(24):1850-1852

[14] Lawson WD, Nielsen S, Putley EH, Young AS. Preparation and properties of HgTe and mixed crystals of HgTe-CdTe. Journal of Physics and Chemistry of Solids. 1959;**9**(3-4):325-329

[15] Rogalski A. HgCdTe infrared detector material: History, status and outlook. Reports on Progress in Physics. 2005;**68**(10):2267

[16] Arias JM, Pasko JG, Zandian M, Shin SH, Williams GM, Bubulac LO, et al. MBE HgCdTe heterostructure p-on-n planar infrared photodiodes. Journal of Electronic Materials. 1993;**22**(8):1049-1053

[17] Destefanis G, Chamonal JP. Large improvement in HgCdTe photovoltaic detector performances at LETI. Journal of Electronic Materials. 1993;**22**(8):1027-1032

[18] Smith EPG, Pham LT, Venzor GM, Norton EM, Newton MD, Goetz PM, et al. HgCdTe focal plane arrays for dual-color mid-and long-wavelength infrared detection. Journal of Electronic Materials. 2004;**33**(6):509-516

[19] Smith JS, Chiu LC, Margalit S, Yariv A, Cho AY. A new infrared detector using electron emission from multiple quantum wells. Journal of Vacuum Science & Technology, B: Microelectronics Processing and Phenomena. 1983;**1**(2):376-378

[20] Esaki L, Sakaki H. New photoconductor. IBM Technical Disclosure Bulletin. 1977;**20**:2 456-2457

[21] Chiu LC, Smith JS, Margalit S, Yariv A, Cho AY. Application of internal photoemission from quantum-well and heterojunction superlattices to infrared photodetectors. Infrared Physics. 1983;**23**(2):93-97

[22] Coon DD, Karunasiri RPG. New mode of IR detection using quantum wells. Applied Physics Letters. 1984;**45**(6):649-651

[23] Levine BF, Choi KK, Bethea CG, Walker J, Malik RJ. New 10 μm infrared detector using intersubband absorption in resonant tunneling GaAlAs superlattices. Applied Physics Letters. 1987;**50**(16):1092-1094

[24] Levine BF. Quantum-well infrared photodetectors. Journal of Applied Physics. 1993; **74**(8):R1-R81

[25] Henini M. QWIPs enhance infrared detection. III-Vs Review. 1998;**11**(3):30-34

[26] Beeler M, Trichas E, Monroy E. III-nitride semiconductors for intersubband optoelectronics: A review. Semiconductor Science and Technology. 2013;**28**(7):074022

[27] Liu HC, Capasso F, editors. Intersubband Transitions in Quantum Wells: Physics and Device Applications I and II. San Diego, CA: Academic; 2000

[28] Odoh EO, Njapba AS. A review of semiconductor quantum well devices. Advances in Physics Theories and Applications. 2015;**46**:26-32

[29] Gunapala SD, Bandara SV, Liu JK, Hill CJ, Rafol SB, Mumolo JM, et al. 1024×1024 pixel mid-wavelength and long-wavelength infrared QWIP focal plane arrays for imaging applications. Semiconductor Science and Technology. 2005;**20**(5):473

[30] Gunapala SD, Bandara SV, Liu JK, Luong EM, Stetson N, Shott CA, et al. Long-wavelength 256/spl times/256 GaAs/AlGaAs quantum well infrared photodetector (QWIP) palm-size camera. IEEE Transactions on Electron Devices. 2000;**47**(2):326-332

[31] Dhingra M, Shankar A, Tiwari BB. A review on quantum well structures in photonic devices for enhanced speed and span of the transmission network. Indian Journal of Physics. 2010;**84**(8):1031-1037

[32] Andersson JY, Lundqvist L, Paska ZF. Quantum efficiency enhancement of AlGaAs/GaAs quantum well infrared detectors using a waveguide with a grating coupler. Applied Physics Letters. 1991;**58**(20):2264-2266

[33] Sarusi G, Levine BF, Pearton SJ, Bandara KMS, Leibenguth RE, Andersson JY. Optimization of two dimensional gratings for very long wavelength quantum well infrared photodetectors. Journal of Applied Physics. 1994;**76**(9):4989-4994

[34] Choi KK, Lin CH, Leung KM, Tamir T, Mao J, Tsui DC, et al. Broadband and narrow band light coupling for QWIPs. Infrared Physics & Technology. 2003;**44**(5-6):309-324

[35] Levine BF, Gunapala SD, Kuo JM, Pei SS, Hui S. Normal incidence hole intersubband absorption long wavelength GaAs/AlxGa1−xAs quantum well infrared photodetectors. Applied Physics Letters. 1991;**59**(15):1864-1866

[36] Gunapala SD, Bandara SV, Liu JK, Rafol SB, Mumolo JM, Shott CA, et al. 640 × 512 pixel narrow-band, four-band, and broad-band quantum well infrared photodetector focal plane arrays. Infrared Physics & Technology. 2003;**44**(5-6):411-425

[37] Gunapala SD, Bandara SV, Liu JK, Rafol SB, Mutnolo JM. 640/spl times/512 pixel long-wavelength infrared narrowband, multiband, and broadband QWIP focal plane arrays. IEEE Transactions on Electron Devices. 2003;**50**(12):2353-2360

[38] Kato K. Ultrawide-band/high-frequency photodetectors. IEEE Transactions on Microwave Theory and Techniques. 1999;**47**(7):1265-1281

[39] Umbach A, Trommer D, Steingrüber R, Seeger A, Ebert W, Unterbörsch G. High-speed, high-power 1.55 μm photodetectors. Optical and Quantum Electronics. 2001; **33**(7-10):1101-1112

[40] Bowers JE, Burrus CA. High-speed zero-bias waveguide photodetectors. Electronics Letters. 1986;**22**:905-906

[41] Kato K, Hata S, Kozen A, Yoshida J. High-efficiency waveguide InGaAs *p-i-n* photodiode with bandwidth of greater than 40 GHz. In: OFC'91, 1991

[42] Wake D, Spooner TP, Perrin SD, Henning ID. 50 GHz InGaAs edge-coupled pin photodetector. Electronics Letters. 1991;**27**:1073-1075

[43] Wanlin G, Giraudet L, Praseuth JP, Miras A, Legros E. High responsivity side illuminated AlGaInAs pin photodiode for 40 Gbit/s-40 GHz applications. In: ECOC'97. Vol. 2. 1997. pp. 37-40

[44] Unterborsch G, Umbach A, Trommer D, Mekonnen GG. 70 GHz long-wavelength photodetector. In: 11th International Conference on Integrated Optics and Optical Fibre Communications,, and 23rd European Conference on Optical Communications (Conf. Publ. No.: 448). Vol. 2; IET. 1997. pp. 25-28

[45] Bach HG, Beling A, Mekonnen GG, Kunkel R, Schmidt D, Ebert W, et al. InP-based waveguide-integrated photodetector with 100-GHz bandwidth. IEEE Journal of Selected Topics in Quantum Electronics. 2004;**10**(4):668-672

[46] Kato K, Yoshida J. Ultrawide-bandwidth 1.55-m waveguide *p-i-n* photodiode. Proceedings of SPIE-The International Society for Optical Engineering. 1994;**2149**:312-319

[47] Kato K, Kozen A, Muramoto Y, Itaya Y, Nagatsuma T, Yaita M. 110-GHz, 50% efficiency mushroom-mesa waveguide *p-i-n* photodiode for a 1.55-mm wavelength. IEEE Photonics Technology Letters. 1994;**6**:719-721

[48] Taylor HF, Eknoyan O, Park CS, Choi KN, Chang K. Traveling wave photodetectors. Proceedings of SPIE-The International Society for Optical Engineering. 1990;**1217**:59-63

[49] Heitala VM, Vawter GA. A large-bandwidth high-quantum-efficiency traveling-wave photodetector based on a slow-wave coplanar transmission line. In: Prog. Electromagnetics Res. Symp.; July 1991; Cambridge, MA

[50] Giboney K, Nagarajan R, Reynolds T, Allen S, Mirin R, Rodwell M, Bowers J. 172 GHz, 42% quantum efficiency *p-i-n* travelling-wave photodetector. In: 52nd Annu. Device Res. Conf.; June 1194; Boulder, CO. Vol. VIA-9

[51] Giboney KS, Rodwell JW, Bowers JE. Traveling-wave photodetector theory. IEEE Transactions on Microwave Theory and Techniques. 1997;**45**(8):1310-1319

[52] Shi JW, Gan KG, Chiu YJ, Chen YH, Sun CK, Yang YJ, et al. Metal-semiconductor-metal traveling-wave photodetectors. IEEE Photonics Technology Letters. 2001;**13**(6):623-625

[53] Lin LY, Wu MC, Itoh T, Vang TA, Muller RE, Sivco DL, et al. High-power high-speed photodetectors design, analysis, and experimental demonstration. IEEE Transactions on Microwave Theory and Techniques. 1997;**45**:1320-1331

[54] Giboney KS, Nagarajan RL, Reynolds TE, Allen ST, Mirin RP, Rodwell MJ, et al. Travelling-wave photodetectors with 172-GHz bandwidth and 76-GHz bandwidth-efficiency product. IEEE Photonics Technology Letters. 1995;**7**(4):412-414

[55] Ünlü MS, Strite S. Resonant cavity enhanced photonic devices. Journal of Applied Physics. 1995;**78**(2):607-639

[56] Abaeiani G, Ahmadi V, Saghafi K. Design and analysis of resonant cavity enhanced-waveguide photodetectors for microwave photonics applications. IEEE Photonics Technology Letters. 2006;**18**(15):1597-1599

[57] Cho SY, Jokerst NM. Integrated thin film photodetectors with vertically coupled microring resonators for chip scale spectral analysis. Applied Physics Letters. 2007;**90**(10):101105

[58] Chen L, Lipson M. Ultra-low capacitance and high speed germanium photodetectors on silicon. Optics Express. 2009;**17**(10):7901-7906

[59] Ackert JJ, Fiorentino M, Logan DF, Beausoleil R, Jessop PE, Knights AP. Silicon-on-insulator microring resonator defect-based photodetector with 3.5-GHz bandwidth. Journal of Nanophotonics. 2011;**5**(1):059507

[60] Zohar M, Auslender M, Hava S, Faraone L. Resonance cavity enhanced midinfrared photodetectors employing subwavelength grating. In: 11th International Conference on Numerical Simulation of Optoelectronic Devices (NUSOD); September 2011; IEEE; 2011. pp. 25-26

[61] Duan X, Huang Y, Ren X, Shang Y, Fan X, Hu F. High-efficiency InGaAs/InP photodetector incorporating SOI-based concentric circular subwavelength gratings. IEEE Photonics Technology Letters. 2012;**24**(10):863-865

[62] Lai KW, Lee YS, Fu YJ, Lin SD. Selecting detection wavelength of resonant cavity-enhanced photodetectors by guided-mode resonance reflectors. Optics Express. 2012;**20**(4):3572-3579

[63] Ghioni M, Gulinatti A, Rech I, Zappa F, Cova S. Progress in silicon single-photon avalanche diodes. IEEE Journal of Selected Topics in Quantum Electronics. 2007;**13**(4):852-862

[64] Itzler MA, Jiang X, Entwistle M, Slomkowski K, Tosi A, Acerbi F, et al. Advances in InGaAsP-based avalanche diode single photon detectors. Journal of Modern Optics. 2011;**58**(3-4):174-200

[65] Ghioni M, Armellini G, Maccagnani P, Rech I, Emsley MK, Unlu MS. Resonant-cavity-enhanced single-photon avalanche diodes on reflecting silicon substrates. IEEE Photonics Technology Letters. 2008;**20**(6):413-415

[66] Ghioni M, Armellini G, Maccagnani P, Rech I, Emsley MK, Ünlü MS. Resonant-cavity-enhanced single photon avalanche diodes on double silicon-on-insulator substrates. Journal of Modern Optics. 2009;**56**(2-3):309-316

[67] Cansizoglu H, Devine EP, Gao Y, Ghandiparsi S, Yamada T, Elrefaie AF, et al. A new paradigm in high-speed and high-efficiency silicon photodiodes for communication—Part I: Enhancing photon-material interactions via low-dimensional structures. IEEE Transactions on Electron Devices. 2018;**65**(2):372-381

[68] Cansizoglu H, Elrefaie AF, Bartolo-Perez C, Yamada T, Gao Y, Mayet AS, et al. A new paradigm in high-speed and high-efficiency silicon photodiodes for communication—Part II: Device and VLSI integration challenges for low-dimensional structures. IEEE Transactions on Electron Devices. 2018;**65**(2):382-391

[69] Zang K, Jiang X, Huo Y, Ding X, Morea M, Chen X, et al. Silicon single-photon avalanche diodes with nano-structured light trapping. Nature Communications. 2017;**8**(1):628

[70] Gao Y, Cansizoglu H, Polat KG, Ghandiparsi S, Kaya A, Mamtaz HH, et al. Photon-trapping microstructures enable high-speed high-efficiency silicon photodiodes. Nature Photonics. 2017;**11**(5):301

[71] Ishi T, Fujikata J, Makita K, Baba T, Ohashi K. Si nanophotodiode with a surface plasmon antenna. Japanese Journal of Applied Physics. 2005;**44**(3L):L364

[72] Goykhman I, Sassi U, Desiatov B, Mazurski N, Milana S, de Fazio D, et al. On-chip integrated, silicon-graphene plasmonic Schottky photodetector with high responsivity and avalanche photogain. Nano Letters. 2016;**16**(5):3005-3013

[73] Levy U, Grajower M, Goncalves PAD, Mortensen NA, Khurgin JB. Plasmonic silicon Schottky photodetectors: The physics behind graphene enhanced internal photoemission. APL Photonics. 2017;**2**(2):026103

[74] Muehlbrandt S, Melikyan A, Harter T, Köhnle K, Muslija A, Vincze P, et al. Silicon-plasmonic internal-photoemission detector for 40 Gbit/s data reception. Optica. 2016;**3**(7):741-747

[75] Soci C, Zhang A, Bao XY, Kim H, Lo Y, Wang D. Nanowire photodetectors. Journal of Nanoscience and Nanotechnology. 2010;**10**(3):1430-1449

[76] Yan C, Lee PS. Recent progresses in improving nanowire photodetector performances. Science of Advanced Materials. 2012;**4**(2):241-253

[77] Logeeswaran VJ, Oh J, Nayak AP, Katzenmeyer AM, Gilchrist KH, Grego S, et al. A perspective on nanowire photodetectors: Current status, future challenges, and opportunities. IEEE Journal of Selected Topics in Quantum Electronics. 2011;**17**(4):1002-1032

[78] Zhai T, Fang X, Liao M, Xu X, Zeng H, Yoshio B, et al. A comprehensive review of one-dimensional metal-oxide nanostructure photodetectors. Sensors. 2009;**9**(8):6504-6529

[79] LaPierre RR, Robson M, Azizur-Rahman KM, Kuyanov P. A review of III-V nanowire infrared photodetectors and sensors. Journal of Physics D: Applied Physics. 2017;**50**(12):123001

[80] Logeeswaran VJ, Sarkar A, Islam MS, Kobayashi NP, Straznicky J, Li X, et al. A 14-ps full width at half maximum high-speed photoconductor fabricated with intersecting InP nanowires on an amorphous surface. Applied Physics A. 2008;**91**(1):1-5

[81] Gallo EM, Chen G, Currie M, McGuckin T, Prete P, Lovergine N, et al. Picosecond response times in GaAs/AlGaAs core/shell nanowire-based photodetectors. Applied Physics Letters. 2011;**98**(24):241113

[82] Farrell AC, Senanayake P, Hung CH, El-Howayek G, Rajagopal A, Currie M, et al. Plasmonic field confinement for separate absorption-multiplication in InGaAs nanopillar avalanche photodiodes. Scientific Reports. 2015;**5**:17580

Chapter 8

Linear CCD-Based Spectrometry Using Either an ASIC or FPGA Design Methodology

Richard Ocaya

Additional information is available at the end of the chapter

http://dx.doi.org/10.5772/intechopen.81654

Abstract

At room temperature, high-responsivity charge-coupled devices (CCD) comprising arrays of several thousand linear photodiodes are readily available. These sensors are capable of ultraviolet to near infrared wavelengths sensing with detecting resolutions of up to 24 dots per millimeter. Their applicability in novel spectrometry applications has been demonstrated. However, the complexity of their timing, image acquisition, and processing necessitates sophisticated peripheral circuitry for viable output. In this chapter, we outline the application specifications for a versatile spectrometer that is reliant on a field programmable gate array (FPGA) automation. The sustained throughput is 1.23 gigabit per second 8-bit color readout rate. This approach is attractive because the final FPGA design may be reconfigured readily to a single, branded, application-specific integrated circuit (ASIC) to drive a wider range of linear CCDs on the market. This is advantageous for rapid development and deployment of the spectrometer instrument.

Keywords: linear image sensing, FPGA, ASIC, image acquisition, high-speed processing

1. Introduction

The proliferation of imaging devices in many applications today is due to the significant technological progress that has occurred over the last few decades in the area of image sensing, particularly with respect to charge-coupled device (CCD) sensors. Today, they are found everywhere from document line-scanned imaging to high-definition planar image acquisition, thereby covering a wide variety of applications. The interest to use CCDs in serious scientific instruments arose from the advances in the area of high-sensitivity, large-area and low-noise CCDs. These CCDs began to routinely provide a high quantum efficiency (QE) figure for each of the millions of pixels, low-noise readout over wide spectral and dynamic ranges. One factor that contributed to these advances is that solutions were given to many of the problems associated with defect states of the semiconductor substrates on which the devices were built. Defects lead to low charge transfer efficiencies due to their lossy nature. This made it difficult to fabricate large format image sensors. Various other advancements allowed the reduction of readout noise, thereby improving the photodiode sensitivity even further [1]. Of particular note is the electron multiplication scheme that provided on-chip gain with a net effect that is analogous to the operation of the photomultiplier tube (PMT). The typical CCD imager consists of a coordinated collection of individual photodiodes that can range from a few hundred pixels for computer optical mouse position encoding to high-speed high-definition imaging [2]. For many applications, the CCDs consist of primary color composite arrays that have image resolutions approaching 1200 dots per inch, excellent low-lux sensitivity. In general, devices with high QE and high detection sensitivity over wide wavelengths are fast-approaching performances nearly equal to the traditional photomultiplier tube, with an added bonus of having built-in control electronics that require only correct timing and serially orientated data acquisition [3–7]. The vast number of individual photosensors and serial readout for the typical CCD sensor necessitates a rather complex layout of additional components that are peripheral to the CCD for the purposes of control phasing and the actual data acquisition. CCDs have been successfully applied to many scientific applications, such as the cryogenically cooled CCDs in astronomy applications, in spectroscopy, and in education [8–14]. The operational principle of a CCD is simple in practice, namely that the individual photosensors are synchronously and serially clocked output. This poses formidable implementational challenges due to the complexity of biasing, timing, and control on any CCD device, for a usable intensity output data stream with high bandwidth images. According to the Nyquist sampling theorem, acquisition must therefore proceed at least twice the image bandwidth [15]. In practice, the analog-to-digital converter (ADC) must operate faster than the Nyquist frequency because of the mandatory intermediate processes in the sequence of acquisition, from initialization to acquired image transmission to data terminal equipment (DTE). There are a number of usable approaches to the implementation of the sequences in the image acquisition from a CCD device. The first is the basic microcontroller- or microprocessor-based approach. The second is based on a field programmable gate array (FPGA), and the third is based on an application-specific integrated circuit (ASIC)-based design. The third strategy is based on a custom-designed hardware application, known as an application-specific integrated circuit (ASIC). The ASIC approach offers a competing strategy to FPGA that, despite its generally higher speed and throughput, is likely to better suit larger corporations that have the resources to develop and manufacture specific integrated circuit designs. The needs for mass deployment of an ASIC require justification, such as the range and the volume of the final units required. Where the ASIC is the desired implementation, it is more likely that an FPGA will be used in the developmental stage for design and validation before a custom ASIC can be developed. The cores of such an FPGA-derived ASIC can have provisions for upgrading, such as externally connected dual data rate (DDR) block memories and secure device cards (SD cards) for storage upgrading, user-specifiable display options, USB communications, and other peripheral technological features. Generally, it has become trivial to configure the internal complex logic blocks into a user-defined architecture with optimized memory and execution speed in a manner that is less reliant on additional peripherals and firmware.

An alternative solid-state imaging technology to the CCD is commonly referred to as active pixel sensors (APSs), which are based on complementary metal oxide semiconductors (CMOSs). The allure of the CMOS transistor is its low form factor (FF) on the semiconductor substrate in comparison with bipolar devices. In fact, advances in CMOS fabrication techniques are largely responsible for the microprocessor revolution because they are more suitable for large-scale integration (LSI). Therefore, APS devices have the primary advantage of having higher pixel densities; hence, these devices have larger pixel arrays and wider photon collection areas. The simplicity of the APS sensing mechanism, which comprises single photodiode and at most three transistors, is vastly reduced in comparison with that of the CCD. A detection element in an APS array essentially follows a randomized access, row-column addressing protocol. The CCD element, on the other hand, relies on a sequential conveyance of charge, leading to a framing approach to image recovery. The general consensus appears to be that CMOS sensors suitable for scientific instrumentation still have much room for improvement with respect to QE, in spite of having higher image access speeds [1, 16, 17]. The low QE figures for APS devices stem from what is referred to as the "fill factor," which is a measure of the actual detection area to the entire area of the APS element. Although slow, CCDs have a high fill-factor and a large full-well capacity which makes them suitable for astrological imaging. The need to improve the QE figure has been the subject of active research and development, which has led to the attainment of CMOS performance at par with CCDs. Furthermore, unlike CMOS arrays, CCD pixels are not amenable to avalanche-gain enhancement at a given detection site, frequency/phase lock-in, and do not benefit from local pixel amplifiers to improve the signal-to-noise (SNR) figure. Also, CMOS benefits significantly from time-correlated imaging, global shutter synchronization, photon counting, and 0.1–0.5e- (sub-electron) RMS readout noise levels. The definition of color at pixel level through color filtering by using different p-n junction layers is easier in CMOS. Hybridized devices comprising CCD and CMOS that capitalize on the desirable aspects of each technology are also now being developed [18, 19]. A detailed comparison between CCD and CMOS imagers is beyond our present scope. Holst and Lomheim [20], Janesick et al. [21, 22], and others provide a good review of the two alternatives.

We begin by discussing the design methodologies that might be considered when designing a spectrometer for scientific applications, such as wavelength resolvable imaging using a linear CCD. Then, a typical FPGA spectrometer design based on an exemplary CCD is described after laying out the rationale for why the FPGA is preferred for the design. The approach taken in this work is to present a proof-of-concept for the overall system functionality rather than as a final implementation. The FPGA is therefore being used merely for design verification with the intention to re-synthesize onto a purely ASIC system. However, it must be noted that the proprietary nature of the Xilinx IP cannot be synthesized without a license onto non-Xilinx ASICs processes from various manufacturers, such as the Taiwan Semiconductor Manufacturing Company (TMSC) or Global foundries. This would then necessitate the design of equivalent codes for FIFOs and other IPs before implementation using ASICs.

2. FPGA or ASIC?

For high-volume productions, the lower unit cost of an ASIC has generally made it attractive in comparison to FPGAs. FPGAs are now widely recognized as using leading edge technologies in order to obtain the same system performance as an ASIC in older technologies; an increasing number of FPGA-based systems are routinely being converted to ASICs. However, the appeal of ASICs transcends the issue of cost alone. ASIC systems generally have significantly much lower power consumption, and this is a bonus in battery-operated mobile devices such as cameras and mobile phones. Also, the hard coding of the logic in an ASIC leads to more secure and reliable systems by making it virtually impossible to reprogram the device. This reliability makes ASICs the obvious choice over FPGAs for critical applications. Over the past two decades, there has been a gradual movement toward the development and application of femtosecond time-resolved spectroscopy (TRS) in many areas of measurement [23–29]. Such imagers are used quite extensively with CCD and CMOS cameras to give high readout rates. Such fast performance places a significant demand on the control system. In TRS, fixed position measurements are correlated at fixed photon transit times. This is largely due to the increasing number of pixels and resolution of these cameras [30–34]. In general, the processing requirements of temporally demanding applications that maintain high sensitivity within a single package tend to transcend the capability of commercially sourced off-the-shelf components. ASICs have amicably risen to the challenge and are finding increasing usage for such applications, particularly because they do away with the need for a dedicated, high-throughput ADC altogether. Thus, these developments mean that printed circuit board level design practices of ADCs are no longer needed. The characterization of the ADC becomes critical in such applications, but ASICs and FPGAs are natively suited for speed optimization during the design phases. High-input bandwidth ADCs with sampling frequencies of over 300 Ms./s that digitize events with femtosecond resolution and low crosstalk are now commonly implemented using ASICs [35–37] as well as FPGAs [38]. Such ASICs are found in critical experiments such as the Large Hadron Collider (LHC) [39], space [1], organic, and biomedical applications [19–27].

2.1. The FPGA design flow

Figure 1 depicts the design steps in a typical FPGA-based design process. In general, FPGAs provide reduced design time and bug fixes due to faulted design logic since the logic design step in customary ASIC designs is absent. The verification of deep-submicron placement issues, particularly with respect to performance issues of heat removal and speed, is easily achieved. The prototype designs can be verified virtually instantly during development as many times as necessary by simply downloading the design into the development test bed. The disadvantage is that a given FPGA design relies heavily on the programmer's abilities to write efficient FPGA code. The performance of the design therefore somewhat relies on developer's ability. As with ASICs, there is a clear need to optimize the hardware instance.

2.2. The ASIC design flow

Figure 2 depicts the design steps in a typical ASIC-based design process. ASIC tools are generally script driven, unlike FPGAs. In the development process, FPGA system designers are

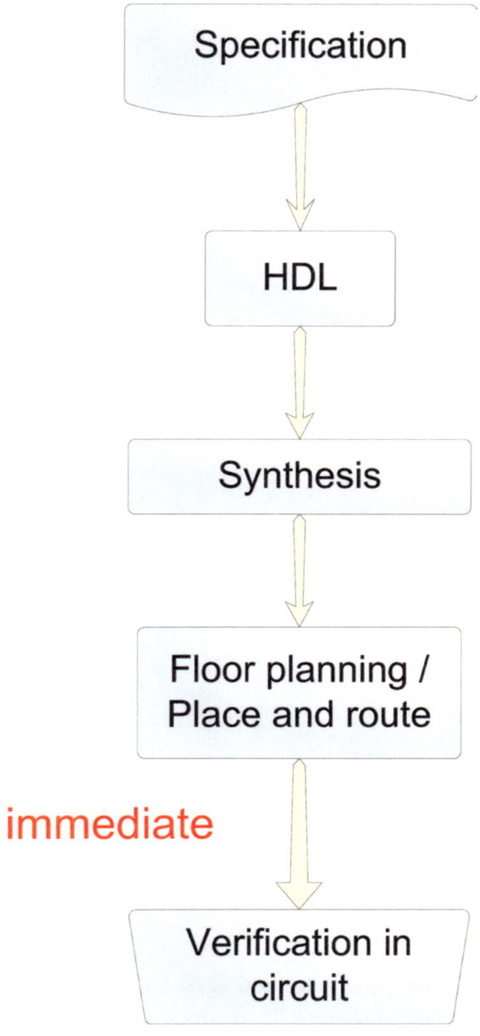

Figure 1. Typical FPGA design flow.

increasingly reliant on graphical user interfaces (GUIs), although the very high speed integrated circuit (VHSIC) hardware definition logic (VHDL) is inherently script driven. In ASIC design, postsynthesis analyses of the timing and functional equivalences are the responsibility of the system designer before prototyping. The effects of deep-micron logic element placements need careful appraisal by the designer, unlike in FPGAs where they are routinely part of the design verification step. Designing using FPGAs is therefore associated with fast turnaround.

2.3. A typical FPGA application in CCD-based spectroscopy

Figure 3 shows the schematic of a typical linear CCD application in which the light focused on the CCD by a lens array is derived from a reflective or transmissive grating after striking

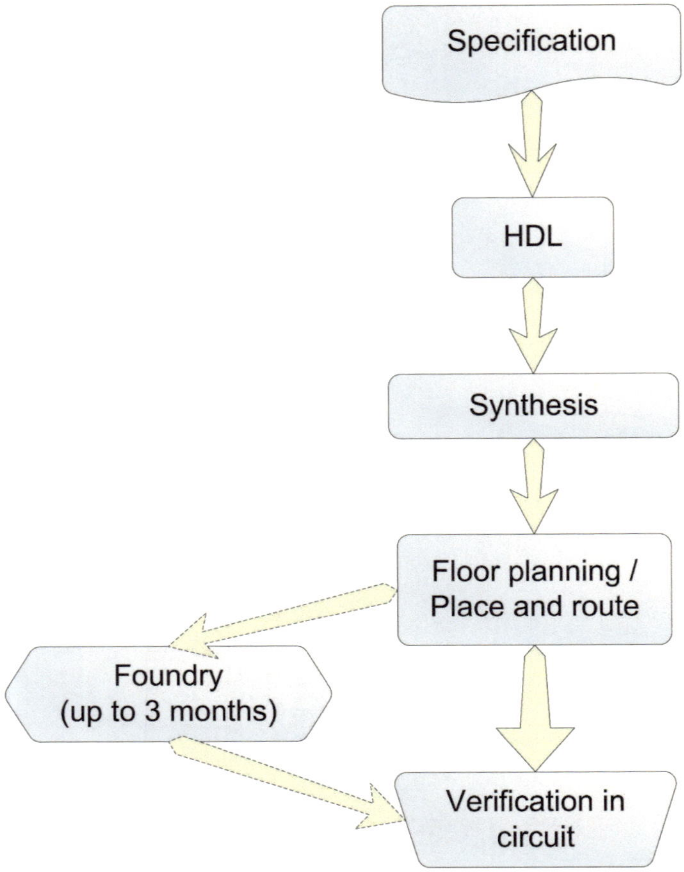

Figure 2. Typical ASIC design flow.

a carefully positioned sample being characterized [14]. The general elements in the layout are identifiable in other specific implementations. The output of the grating is proportional to wavelength and manifest as angularly dispersed alternating zones of high and low light intensity with a linear resolution in terms of spectral, that is, wavelength spread or range per unit length [2].

1. **Figure 3** shows the general schematic of the image forming optics. A real but inverted image, shown as x''_λ, is required to fall onto the photosensitive areas of the CCD for detection. This image is a magnification of the virtual object x'_λ, which is itself the result of net interference of the light source as it passes through the grating. It is easy to show that the spread of wavelength along the total detection array length of the CCD is distance resolved. Thus, assuming that the achromatic reduction lens has image magnification m, where generally $m < 1$, then the lengthwise spread of the image along the CCD's photosensors is:

$$x''_\lambda = \alpha\lambda, \tag{1}$$

where $\alpha = mL/d$.

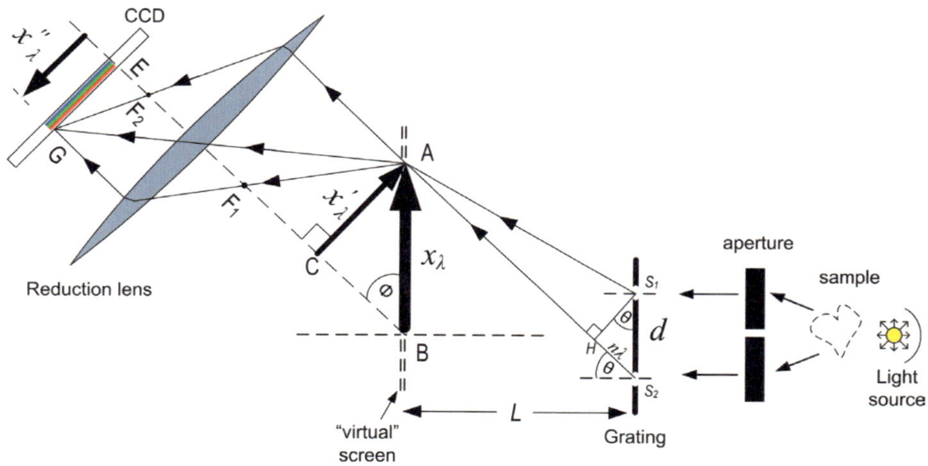

Figure 3. Layout of the image-forming arrangement typical in a scientific spectrometer.

Therefore, a bright spot located at the position x_λ'' will be at wavelength λ and an intensity measured at the corresponding photodiode. Hence, a direct readout of the CCD intensity data is possible. By utilizing a standardized light source, the voltage output of the CCD can be calibrated in lux; although this is not necessary in a spectroscopic application where knowing the spectral character and the relative peak intensities is more important than actual intensity, it is normal to express the output count in arbitrary units (au). Therefore, a plot of the CCD sensor output, which is proportional to the incident intensity, versus wavelength suffices for many applications. In our particular case, the CCD output is measured in volts. The CCD output is read out sequentially in step with synchronization pulses. The hardware or software that receives the CCD output should therefore be designed to interpret that the CCD output as being within a well-defined frame. **Figure 4** shows the timing and control signals that are typical in linear CCDs for the detection of three colors. The waveform-labeled OS (serial output) shows the intensity variations relative to the sample and hold (SH), clocking phases (Φ1 and Φ2), and the register shift (RS) signals.

Figure 4 shows the complexity of the signals required to drive a typical CCD in even the most cursory application, without data acquisition in any form. These signals are derived from an actual CCD, the TCD2557, which henceforth we shall refer to as the exemplary application. The complexity of the additional conditioning circuitry can be appreciated by considering the following scenario. The exemplary CCD has 5415 elements that make up the readout frame, but 5340 actual photosensors. The remaining 75 elements are dummy sensors, 64 presensor blocks and 11 postsensor blocks. The CCD requires dual-phase clock signals (Φ1 and Φ2) which, if run at 1.25 MHz ($\tau = 0.8\mu s$), define the shortest frame time of $5415\tau = 4.332$ ms. The needs of additional processing cause timing overheads beyond the needs to construct the frame. USB transfers and other processes are nonparallel and consume added real time. Between the three afore-mentioned design strategies that would realize a practical spectrometer that is designed around the exemplary linear CCD, the microcontroller approach is the least efficient for the following reason. If the ADC presumably uses an 8-bit successive-approximation, ADC converter

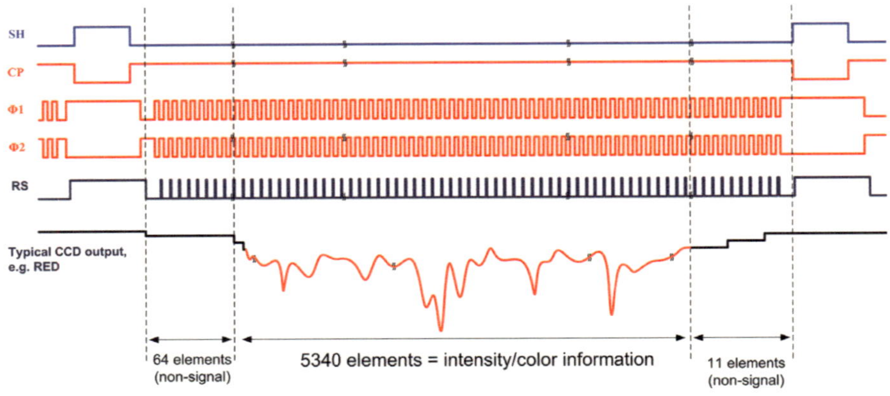

Figure 4. Timing diagrams that lead to the generation of CCD output.

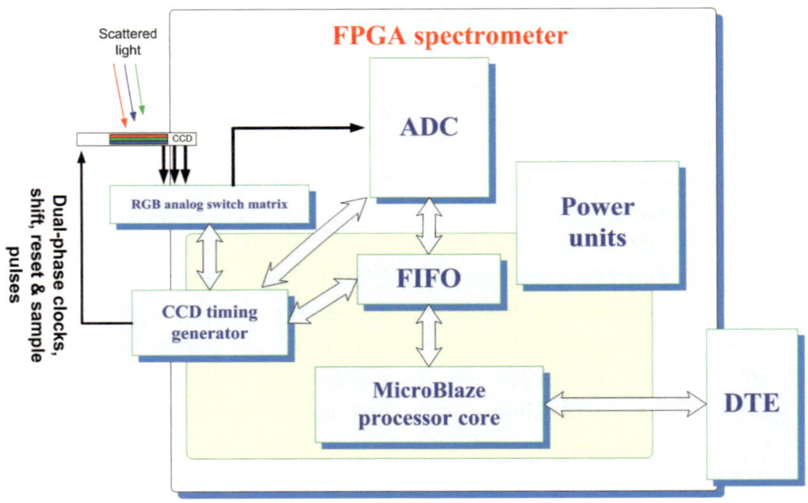

Figure 5. Block diagram of an FPGA-driven CCD-based spectrometer.

running at 1 MHz (i.e., $\tau = 1\mu s$), then each photodiode level is converted to at least $2^8\tau = 2.56$ ms. It is likely that an associated temporary memory for $M = 8192$ $(4192 < (M = 5340) < 8192)$ bytes will be required. The operation of such a memory will be such that at the end of each conversion the memory is written with the ADC result. All photodiodes are then completely digitized in $5340 \times 2^8\tau \approx 13.67$ s. This time is clearly far too long for a simple 8-bit acquisition. In addition, the firmware-defined timing signals need to be generated and issued to the CCD from the free output lines of the microcontroller, further complicating the design and requiring an extremely fast microcontroller. It is estimated that the operating frequency of such a microcontroller would need to be well in excess of 40 MHz. **Figure 5** shows an FPGA-based spectrometer built around a Xilinx Spartan FPGA. Several specific FPGAs may be used. For reasons of cost and

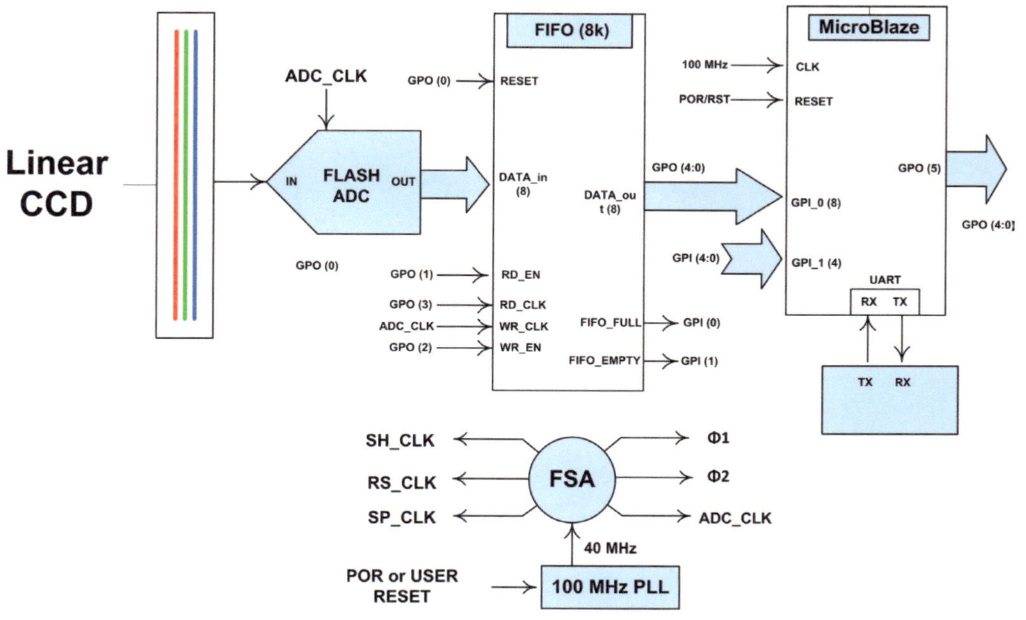

Figure 6. FPGA-based CCD block schematic based on a typical Xilinx FPGA, the Spartan-6LX9. The implementation relies on various intellectual properties (IP) such as MicroBlaze.

availability, the author settled for the Spartan-6 LX9 FPGA device for an actual spectrometer application to drive the TCD2557 CCD [40–43]. The Xilinx software development kit (SDK) allows the organization of the complex logic blocks (CLB) in some of its FPGAs into a powerful, firmware-defined 100-MHz microprocessor referred to as MicroBlaze. With phase-locked loop clock synthesis, MicroBlaze is capable of 400-MHz internal clock speeds. This complex, proprietary engine has all the functionalities of a microprocessor. Furthermore, as with normal microcontrollers, MicroBlaze can be further controlled by a user-defined firmware written in a high-level language such as C/C++. A major advantage of VHDL is that its processes are, by default, parallel and can be made either synchronous or asynchronous. Thus, the generation and handling of the timing and synchronization signals becomes a trivial exercise. **Figure 6** shows the organization of the Spartan-6LX9 to implement the requisite control signals and the connection of the CCD.

The FPGA hardware strategy features a parallel design paradigm that is generally simpler and highly repeatable. The FPGA design thus far is volatile with respect to the power state that is lost when the power to the FPGA is removed. The bitstream file that represents the developed design and defines the application must therefore be stored in an electrically erasable programmable read-only memory (E^2PROM) and reloaded into the FPGA at power on. The role of the intermediate memory is then implemented by defining an 8-kilobyte × 8-bit first-in first-out (FIFO) structure. Using a FIFO buffer allows the readout device to operate at a slower clock rate alongside other parallel processes such as acquisition and USB transfers. In the actual application mentioned above, we have achieved a CCD readout rate of 1.23 gigabit/s per color.

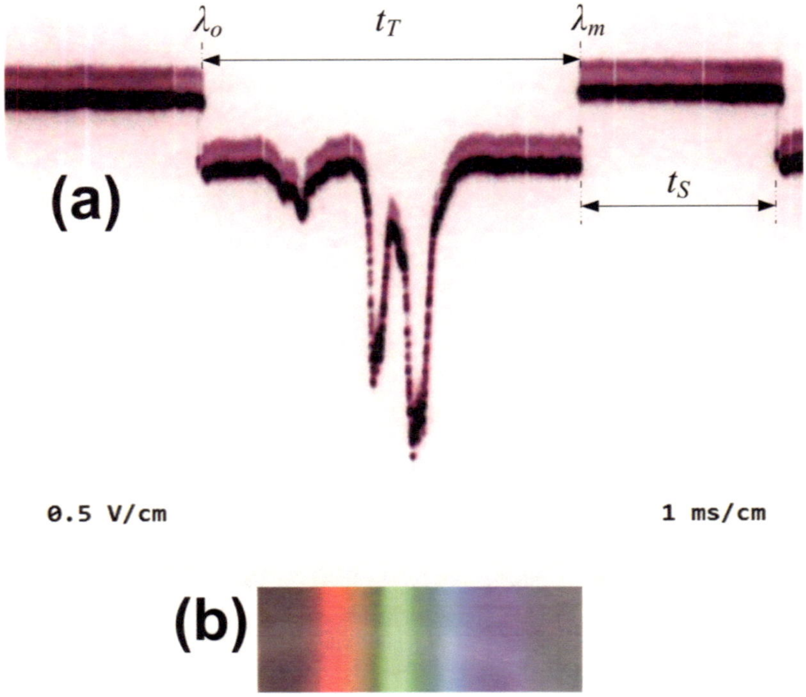

Figure 7. Actual photographs showing (a) the actual TCD2557 CCD output, (b) the CFL lamp spectrum that generates the output in (a).

Figure 7 shows the actual oscilloscope output (a) in response to the diffracted spectrum (b) obtained when a source of illumination was directed at the spectrometer.

2.4. The operational principle

The Spartan-6 MicroBlaze FPGA intellectual property (IP) owned by Xilinx differs from the conventional micro-controller that comes with a fixed architecture and is supported by rigid peripherals in that it permits the arbitrary implementation of high-speed functional blocks from the array of CLBs, clocks, and system management tiles. Their internal structures can be arranged into several virtual, highly parallel microcontrollers and peripherals [41]. The advancements that culminate in MicroBlaze and other IPs are made possible due to the optimization of the logic of these "sixth-generation" FPGAs. The inherent 45-nanometer, copper-interconnected architecture permits devices to be built for speed and low-power consumption. Thus, FPGA-based applications having good cost/power and cost/performance ratios are a reality [42]. There are other FPGAs beside Xilinx made that have well-defined development flow. In the exemplary CCD implementation, we have implemented an efficient system comprising a 2-register, 6-input lookup table (LUT), steering logic, 18 kilobytes of random access memory (RAM), and support for the USB 2.0 standard for communications with the DTE. These features therefore make the FPGA a very good alternative to the microcontroller

http://dx.doi.org/10.5772/intechopen.81654

approach. In contrast to the FPGA approach, there are on the market dedicated CCD management ASICs, such as those developed by the CCD manufacturers themselves. However, the lead time, specificity to CCD, general complexity, and cost are factors to consider during the design of the application. Many ASICs are heuristic, "black box" solutions to complex image acquisition. Such solutions are, for the low volume designs at least, that ASICs can be notoriously closed, proprietary solutions whose innards and general operations are hard to decipher without knowledge of the intellectual property, or some amount of reverse engineering. Therefore, while an ASIC can deliver a blazingly fast and reliable performance in a given application, it can be beyond the means of a low volume production in terms of development cost and ASIC design. In this work, the necessary CCD control signals are derived from a finite state machine (FSM) that operates concurrently with a MicroBlaze core. Huang et al. have described their Xilinx FPGA implementation of a visualization spectrometer [43]. The FSM also synchronizes the 20 MB/s half-flash ADC with the FIFO. It effectively controls the output sequencing of the CCD output into the FIFO structure through the sample-and-hold (SH) signal, thereby allowing FIFO buffer reads by the DTE at rates that are significantly lower than the ADC conversion rate [37, 44, 45]. Much slower DTE interfaces, such as computer soundcard sampled at 44.1 kHz, can readily digitize the intensity stream using a number of third-party software (**Figure 8**).

2.4.1. Emulating a microcontroller strategy using an IP core

The concept of reusable intellectual properties (IP) pioneered by Xilinx in their sixth-generation FPGAs speeds up the design process even further [40]. These are verified, visually presented subunits and instantiation templates that save further development time and effort. The sole requirement of the user is to correctly declare the desired unit and functionality, and then to contextualize the instance of the IP subunit. This approach, which has been employed in our practical instrument with respect to a 100-MHz MicroBlaze core and the FIFO IPs, enhances repeatability and dramatically improves lead-in time. The resulting system is supported by 16 kilobytes of RAM, a general-purpose output register (GPO) of 5-bit width, an 8- and 4-bit general-purpose input (GPI) registers, and communication interfaces consisting of a USB 2.0 port and a universal asynchronous receiver transmitter (UART). The UART is configured for 9600 baud, 8 bits, 1 stop bit, and no parity (9600, 8, N, 1). At present, the system does not process interrupts pertaining to the image acquisition stages.

2.4.2. A description of the FIFO IP core

The implementation of the 8-kilobyte × 8-bit FIFO structure was done requisitioning 18 kilobytes of BRAM and relied on separate clocks for reading and writing to the structure [1]. **Figure 5** shows a diagrammatic depiction of how the FSM [46] scheme was configured to generate the various control and clock signals to drive the CCD and other features in the design. The handshaking and flag-signals handling that are typically associated with FIFO synchronization and flow control by a DTE unit were not used here, for the sake of simplicity and quick realization of a basic spectrometer. The design settled for software flow control instead. Potential data loss can be avoided and the performance maximized by defining a 3-byte threshold that signals a full FIFO buffer. Once detected, the DTE flushes the FIFO. The data transfers to the DTE are

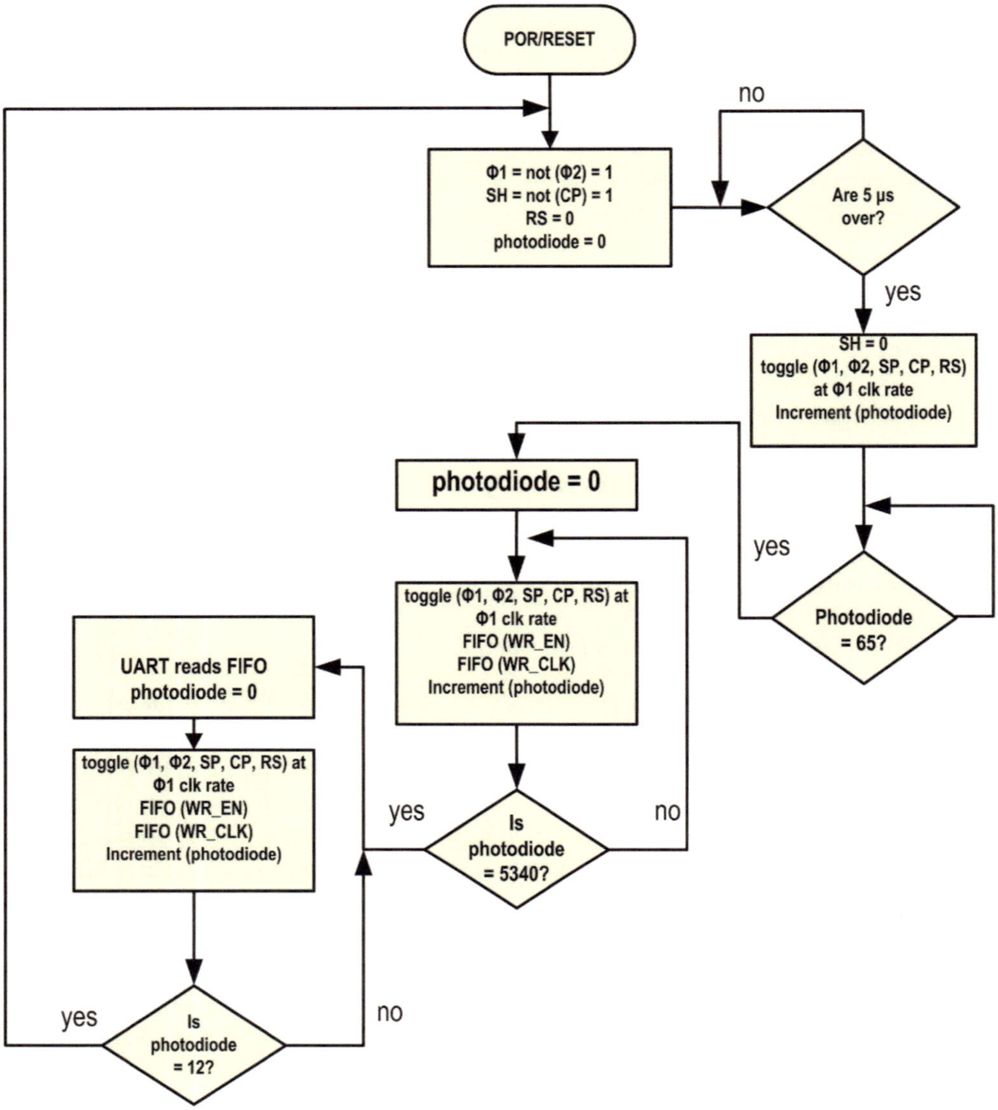

Figure 8. Flowchart of the timing, control, synchronization, and DTE communications using an FPGA or FPGA-derived ASIC strategies.

initiated only upon the detection of the FIFO full state. This approach effectively frees up the DTE to other processing tasks during image acquisition. In **Figure 4**, the rising-edge transitions of the shift-pulse (SP), clamp-pulse (CP), and the photodiode cell reset (RS) produce the analog level-shifting action that defines the CCD operation. In this way, intensity level sensed by a photodiode, which is held as a proportional voltage, is shifted into the next photodiode and eventually out of the CCD analogue output, until all photodiode levels have been shifted out of the frame. All three colors, red, green and blue (RGB), are shifted out in parallel output by the same pulse transitions. The external control of the FIFO by the DTE occurs through the high-level, SDK-developed C/C++ that understands the hardware description, right from the initiation of the design. Upon a request by the DTE, the FPGA sends all of its 8-kilobyte contents to the DTE through the available communication channel. After the reception of the image data by the DTE is completed, the software on the host recomposites the intensity data using a suitable parsing algorithm. At this point, all the intensities and pixels are then matched numerically. The interpretation of these data then yields information about the light that falls on the CCD. This light can be due to the secondary scattering by a sample under characterization.

3. Conclusions

This chapter presents a sufficiently detailed discussion regarding the pros and cons of three approaches in the design of a linear spectrometer instrument that uses a tricolor, CCD image sensor. The completed instrument is capable of being used for serious scientific measurements. The approaches discussed are with respect to the complex timing and signal conditioning requirements to realize a workable output from the CCD. These approaches are (i) microcontroller based, (ii) ASIC based, and (iii) FPGA based. We then described the essential aspects of the design by outlining the CCD and system control signals, the data acquisition and communication needs of the overall instrument. We suggest that the FPGA design approach leads to a high degree of reliability, repeatability for the task since a high performance, for rapid application development. The embedded intellectual properties found on later generations of Xilinx's FPGAs, most notably the MicroBlaze and FIFO IPs, allow rapid application definition, implementation, and testing on a low-cost FPGA. We have discussed an exemplary 20 Ms./s image acquisition system for a spectrometer instrument that allows both USART and USB 2.0 communications with a normal personal computer. The output data stream, comprising intensity versus wavelength format, can be incorporated directly into postprocessing programs. An equation was derived to show how the alignment and calibration of the spectrometer may be done. An aim of this has been to demystify the complexity of the system by outlining in sufficient detail the physics behind image sensing while presenting the overarching challenges that such sensing presents to the acquisition system. The sensing of wavelength-proportional intelligence with a resolution suitable for serious scientific work clearly generates vast amounts of data, from the sensor frontend up until the final acquisition and storage on the DTE, on a time domain that approaches real time. This naturally raises questions about the best acquisition strategy. We evaluate the pros and cons of the FPGA approach versus the ASIC. For a small-scale development, the FPGA provides a quick route to design completion, whereas the ASIC route may be preferable in larger volume productions. This spectrometer has in fact been deployed in thermoluminescence (TL) measurement, photoluminescence (TL), and line scans. The relative ease with which the FPGA was reconfigured for different, actual CCD displays from different manufacturers aptly demonstrates the versatility of the chosen design approach for once-off or low volume products such as this spectrometer. Characterization of the readout rates using a 400-MHz digital storage oscilloscope (DSO) on the TCD2557 CCD produced a sustained figure of 1.23 gigabit/s.

Author details

Richard Ocaya

Address all correspondence to: ocayaro@ufs.ac.za

University of the Free State, Phuthaditjhaba, South Africa

References

[1] Waltham NR, Prydderch M, Mapson-Menard H, Pool P, Harris A. Development of a thinned back-illuminated CMOS active pixel sensor for extreme ultraviolet spectroscopy and imaging in space science. Nuclear Instruments and Methods in Physics Research Section A: Accelerators, Spectrometers, Detectors and Associated Equipment. 2007;**573**(1-2):250-252. DOI: 10.1016/j.nima.2006.10.259

[2] Ocaya RO. A linear CCD spectrometer based on FPGA for light-source characterization. In: Applied Mechanics and Materials, Proceedings of the International Conference on Mechanical Engineering, Industrial Materials and Industrial Electronics, March 2015, London, UK. Vol. 763. 2015. pp. 120-125. DOI: 10.4028/www.scientific.net/AMM.763.120

[3] Wang Y, McCreery RL. Evaluation of a diode laser/charge coupled device spectrometer for near-infrared Raman spectroscopy. Analytical Chemistry. 1989;**61**(23):2647-2651

[4] Williamson JM, Bowling RJ, McCreery RL. Near-infrared Raman spectroscopy with a 783-nm diode laser and CCD array detector. Applied Spectroscopy. 1989;**43**(3):372-375

[5] Turko BT, Albright KL, King NSP, Yates GJ. Gamma ray spectrometer readout with linear CCD sensor. IEEE Transactions on Nuclear Science. 1992;**39**(5):1336-1340

[6] Fu T, Zhao H, Zeng J, Wang Z, Zhong M, Shi C. Improvements to the three-color optical CCD-based pyrometer system. Applied Optics. 2010;**49**(31):5997-6005

[7] Zhang Y, Zhao B, Wang H, Mao W, Zhang X. Technology of near infrared spectroscopy analysis system design based on fixed grating and charge-coupled device. Transactions of the Chinese Society of Agricultural Engineering. 2009;**25**(9):176-181

[8] Deng XQ, Liu J, Wang ZH. Near-infrared spectrometer optical sensor driving and peripheral circuit design. In: Applied Mechanics and Materials, Proceedings of the 2nd International Conference on Industrial Design and Mechanics Power. Scientific.Net; 2013. DOI: 10.4028/www.scientific.net/AMM.437.674

[9] Podoleanu AG, Taplin SR, Webb DJ, Jackson DA. Channeled spectrum display using a CCD array for student laboratory demonstrations. European Journal of Physics. 1994;**15**(5):266-271

[10] Ormachea O, Urquidi O, Casazola D. Development of a portable low-cost LIBS system. In: Proceedings of SPIE—The International Society for Optical Engineering. SPIE Optical Engineering Press. Vol. 8785. 2013. DOI: 10.1117/12.2025507

[11] Pelletier MJ. Raman spectroscopy using an echelle spectrograph with CCD detection. Applied Spectroscopy. 1990;**44**(10):1699-1705

[12] Poulet P, Uhring W, Hanselmann W, Glazenborg R, Nouizi F, Zint V, et al. A time-gated near-infrared spectroscopic imaging device for clinical applications. In: Proceedings of SPIE, Photonic Therapeutics and Diagnostics IX. SPIE Optical Engineering Press. Vol. 8565. 2013. DOI: 10.1117/12.2003671

[13] Yan F, Zhou Y, Zhang M-C, Chen X. Test equipment for relative spectral response of CCD. Spectroscopy and Spectral Analysis. 2013;**33**(10):2865-2868

[14] Xu H, Liu C-M, He Y, Tang H-W, Wu Q-S. Study on the chemiluminescence resonance energy transfer between luminol and fluorescent dyes using a linear CCD spectrometer. Journal of Luminescence. 2010;**130**(10):1872-1879

[15] Shannon CE. Communication in the presence of noise. Proceedings of the Institute of Radio Engineers. 1949;**37**(1):10-21

[16] Li Z, Deen MJ. Towards a portable Raman spectrometer using a concave grating and a time-gated CMOS SPAD. Optics Express. 2014;**22**(15):18736. DOI: 10.1364/oe.22.018736

[17] Dalla Mora A, Tosi A, Zappa F, Cova S, Contini D, Pifferi A, et al. Fast-gated single-photon avalanche diode for wide dynamic range near infrared spectroscopy. IEEE Journal of Selected Topics in Quantum Electronics. 2010;**16**(4):1023-1030

[18] Janesick JR. Charge coupled CMOS and hybrid detector arrays. In: Proceedings of SPIE—The International Society for Optical Engineering. SPIE Optical Engineering Press. Vol. 5167. 2004. DOI: 10.1117/12.511864

[19] Elloumi M, Eric F, Frederic E, Truchetet F, Cathebras G. Measure of the efficiency of CCD realized in CMOS technology: Application to snapshot video sensors. In: Proceedings of SPIE—The International Society for Optical Engineering. SPIE Optical Engineering Press. Vol. 2869. 1997. DOI: 10.1117/12.273348

[20] Holst GC, Lomheim TS. CMOS/CCD sensors and camera systems. Bellingham, Washington USA: SPIE Optical Engineering Press; 2007

[21] Janesick J, Andrews J, Elliott T. Fundamental performance differences between CMOS and CCD imagers: Part I. Proceedings of SPIE. 2006;**6276**:62760M-1-62760M-19

[22] Janesick J, Andrews J, Tower J, Grygon M, Elliott T, Cheng J, et al. Fundamental performance differences between CMOS and CCD imagers part II. Proceedings of SPIE. 2007;**6690**:669003-1-669003-23

[23] Stolow A, Bragg AE, Neumark DM. Femtosecond time-resolved photoelectron spectroscopy. Chemical Reviews. 2004;**104**:1719-1757

[24] Eppink ATJB, Parker DH. Velocity map imaging of ions and electrons using electrostatic lenses: Application in photoelectron and photofragment ion imaging of molecular oxygen. Review of Scientific Instruments. 1997;**68**(9):3477-3484

[25] Heck AJR, Chandler DW. Imaging techniques for the study of chemical reaction dynamics. Annual Review of Physical Chemistry. 1995;**46**(1):335-372

[26] Falus P, Borthwick MA, Mochrie SGJ. Fast CCD camera for X-ray photon correlation spectroscopy and time-resolved X-ray scattering and imaging. Review of Scientific Instruments. 2004;**75**:4383. DOI: 10.1063/1.1808913

[27] Surber E, Sanov AJ. Photoelectron imaging spectroscopy of molecular and cluster anions: CS^{-2} and $OCS^{-}(H_2O)_{1,2}$. The Journal of Chemical Physics. 2002;**116**:5921. DOI: 10.1063/1.1467916

[28] Gruner SM. Charge-coupled device area X-ray detectors. Review of Scientific Instruments. 2002;**73**:2815. DOI: 10.1063/1.1488674

[29] Whitaker BJ. In: Suits AG, Continetti RE, editors. Imaging in Chemical Dynamics. Washington, DC: American Chemical Society; 2000. p. 68

[30] Zarychta K, Tinet E, Azizi L, Avrillier S, Ettori D, Tualle J-M. Time-resolved diffusing wave spectroscopy with a CCD camera. Optics Express. 2007;**18**(16):16289

[31] Ettori D, Zarychta K, Tinet E, Avrillier S, Tualle J-M: Time-resolved measurement of the scattered light with an interferometric method based on the use of a camera, in Diffuse Optical Imaging of Tissue, R. Cubeddu ed., Vol. 6629 of Proceedings of SPIE-OSA Biomedical Optics (Optical Society of America, 2007), (paper 6629_20). Bellingham, Washington USA: SPIE Optical Engineering Press

[32] Patterson MS, Chance B, Wilson BC. Time resolved reflectance and transmittance for the non-invasive measurement of tissue optical properties. Applied Optics. 1989;**28**(12): 2331-2336

[33] Oppenheim AV, Schafer RW, Buck JR. Discrete-time signal processing. In: Prentice-Hall signal processing series, ISBN 0-13-754920-2. 1999

[34] Gross M, Goy P, Forget BC, Atlan M, Ramaz F, Boccara AC, et al. Heterodyne detection of multiply scattered monochromatic light with a multipixel detector. Optics Letters. 2005;**30**(11):1357-1359. DOI: 10.1364/OL.30.001357

[35] Bult K, Buchwald A. An embedded 240mW 10b 50-MS/s CMOS ADC in 1 mm^2. IEEE Journal of Solid-State Circuits. 1997;**32**:1887-1895

[36] Cline D, Gray P. A power optimized 13-b 5 M samples/s pipelined analog-to-digital converter in 1.2 mm CMOS. IEEE Journal of Solid-State Circuits. 1996;**31**(3):443-452

[37] Xu G: Data acquisition for output signal of linear CCD, Applied Mechanics and Materials, 2011—3rd International Conference on Mechanical and Electronics Engineering, 2011. 130-134. p.2056-2059, ICMEE Hefei, China. Code 87364

[38] Menninga H, Favi C, Fishburn MW, Charbon E. A multi-channel, 10ps resolution, FPGA-based TDC with 300MS/s throughput for open-source PET applications. In: IEEE Nuclear Science Symposium Conference Record. Institute of Electrical and Electronic Engineers (IEEE). 2011. pp. 1515-1522. DOI: 10.1109/NSSMIC.2011.6154362

[39] ALICE TPC—Technical Design Report. A Large Ion Collider Experiment: ALICE TPC—Technical Design Report. ISBN 92-9083-155-3. 1999

[40] Spartan-6 FPGA Family. The low-cost programmable silicon foundation for targeted design platforms. Product Brief PN2402-2. Xilinx. 2011

[41] Spartan 6 LX9 Microboard [Internet]. Available from: https://products.avnet.com/opas-data/d120001/medias/docus/1/Xilinx_s6-lx9-microboard-pb020411.pdf [Accessed May 1, 2016]

[42] DS160 Spartan-6 Family Overview [Internet]. Available from: http://www.xilinx.com/support/documentation/data_sheets/ds160.pdf [Accessed May 1, 2016]

[43] Huang XH, Liu HW, Ye LS. Visualization spectrometer based on refit Xilinx FPGA. In: Applied Mechanics and Materials, Proc. Int. Conf. Mechanical Engineering, Industrial Electronics and Information Technology Applications in Industry. Scientific.Net. Vol. 755. 2013. pp. 427-429. DOI: 10.4028/www.scientific.net/AMM.427-429.755

[44] Chengqun C, Yongfeng R, Fang M. Design of a high-speed image data real-time storage and display circuit. Advanced Materials Research. 2014;**912-914**:1222-1227. DOI: 10.4028/www.scientific.net/AMR.912-914.1222

[45] Gao MF, Yang SP. Design of spectral acquisition system based on the linear array CCD ILX554B. In: Applied Mechanics and Materials, Proc. Int. Conf. Mechatronics Engineering and Computing Technology. Scientific.Net. 2014. pp. 1926-1931. DOI: 10.4028/www.scientific.net/AMM.556-562.1926

[46] Richards CL. An easy way to design complex program controllers. Electronics. 1973;**1**:107-113

Avalanche Photodiode Focal Plane Arrays and Their Application to Laser Detection and Ranging

Hai-Zhi Song

Additional information is available at the end of the chapter

http://dx.doi.org/10.5772/intechopen.81294

Abstract

Focal-plane avalanche photodiodes (APDs) are being more and more widely and deeply studied to satisfy the requirement in weak light and single photon imaging. The progresses of this worldwide study, especially the distinctive researches and achievements in Southwest Institute of Technical Physics and University of Electronic Science and Technology of China are reviewed in this chapter. We successfully fabricated up to 64 × 1 linear-mode Si APD arrays, and 32 × 32–64 × 64 Si single-photon avalanche detector (SPAD) arrays, and applied them in Laser Detection and Ranging (LADAR) platforms like driverless vehicles. Also, we developed 32 × 32–64 × 64 InGaAsP/InP SPAD arrays, and constructed three-dimensional imaging LADAR using them. Together with the progresses of other groups and other materials, we see a prospective future for the development and application of focal-plane APDs.

Keywords: avalanche photodiode, focus plane, laser detection and ranging

1. Introduction

Avalanche photodiodes (APDs) have been widely studied and effectively applied in commercial, military, and academic fields [1] for a few decades. Compared with p-i-n photodiodes, APDs provide higher gain, higher sensitivity and lower detection limit [2], so they are mostly well applied in optical communications [3], imaging [4, 5], and single photon detection [6, 7] in recent years. As all-solid-state optoelectronic devices operating at room-temperature or under thermoelectrically-cooled conditions, APDs are scalable to numerous pixels so that they are taking more and more important roles in focal-plane processing and imaging [8]. Owing to the advantages such as internal photoelectric gain, small size, low driving voltages, high efficiency, and fast response, focal-plane APD arrays bring about new three-dimensional (3D) imaging techniques which provide much wealthier and more accurate information for object recognition and identification [9]. Advanced 3D imaging technologies are strongly required in radar systems including laser detection and ranging (LADAR), so the focal-plane APDs and their LADAR applications were widely and deeply studied in recent years [10–20]. For the purpose of more progress in the future, it is necessary to take an overview on the present research and production of APD arrays. Briefly, the most significant progress is made by MIT Lincoln Laboratory. They developed state-of-the art products of Si and InP/InGaAs Geiger-mode focal-plane arrays [10], which have been successfully applied in a few LADAR systems. Princeton Lightwave also succeeded in producing focal-plane single photon avalanche detector (SPAD) arrays and commercializing their single-photon camera based on the SPADs [11]. The research and production of other groups [12–14] may also be valuable as references for future developments. In this chapter, we review the research and application of the focal-plane APDs in Southwest Institute of Technical Physics and University of Electronic Science and Technology of China [15–20]. It includes linear mode Si APD arrays, Si SPAD arrays and InGaAsP/InP SPAD arrays, which have been applied in LADAR systems.

2. Linear-mode Si APDs

The detection of weak light is technically significant in many application fields such as single molecule fluorescence, high-speed quantum cryptography, and infrared detection [21–23]. In all the application fields, APD devices are strongly required to perform photon-counting with high quantum efficiencies, quick optoelectronic response, and low dark counting rates (noise). LADAR imaging systems work in the way of sampling the spatial and/or temporal information of the optical radiation to an array of detectors. Linear-mode (applied bias slightly lower than the breakdown voltage) APDs are often desired by LADAR systems because their dead-time is normally much shorter than that of Geiger-mode (applied bias slightly higher than the breakdown voltage) APDs so that they can measure sequential pulse returns from closely spaced multiple objects. In extreme cases, linear-mode APDs can even detect a few photons or a single photon, which adds an extra dimension to LADAR scene data [21]. Generally, in the near-infrared spectral band, especially at 905 nm, Si APDs might be applied for ultra-weak light detection, and can be used in linear-mode at gains up to about 500 or greater [23]. Therefore, linear-mode Si APD arrays were developed and applied in LADAR systems.

2.1. Fabrication of the linear-mode Si APD chips

A basic linear-mode APD detector, as shown schematically in **Figure 1**, consists of the APD element and the readout integrated circuit (ROIC) [24, 25]. The ROIC is composed of a trans-impedance amplifier (TIA), a stabilivolt source circuit and a comparer. The APD element converts incident light signal into primary photo-generated carriers and photocurrent, then amplifies the resulting photocurrent through internal avalanche gain, i.e. the impact ionization. The TIA converts the amplified APD current into a voltage signal, which is proportional to the total multiplied charge delivered by the APD.

The design and simulation of an APD device were carried out using a full-band Monte Carlo (MC) device model. For each APD geometry, the MC model incorporates realistic band structures [26, 27]. The basic reach-through APD model with separate layers of absorption, charge and multiplication (SACM) is shown in **Figure 2(a)**. In particular, it is important to know that, in general, electrons can be much more ionized than holes in silicon. Electrons rather than holes should be swept by the electric field into the high field region where the multiplication takes place. Thus, there should be a π-type absorbing region of suitable width for absorbing the incident radiation, and the radiation should be able to enter this region with no loss in any n^+-type layer.

The basic design of a reach-through APD consists of a narrow high-field region where the multiplication takes place, with a much wider low field region in which the incoming radiation is absorbed. As schematically shown in **Figure 2(b)**,an avalanche process occurs as the electric field in a p-n junction is higher than the critical field (E_{cr}) at which impact ionization of carrier starts. The electric field in the p-n junction of a Si APD should be some 2–5 × 10^5 V/cm. It should not be more than 10^6 V/cm at which the Zener effect may take place [28–30].

For satisfactory operation of the APD, the high resistivity π-type substrate must be fully depleted by the applied bias voltage. Generally, it works well provided the substrate wafer is not too thick and the required response times are not less than ~10 ns. However, fabrication

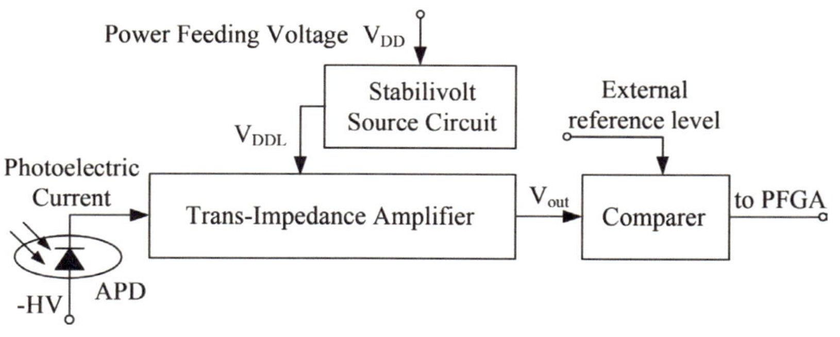

Figure 1. Schematic of the linear-mode Si APD.

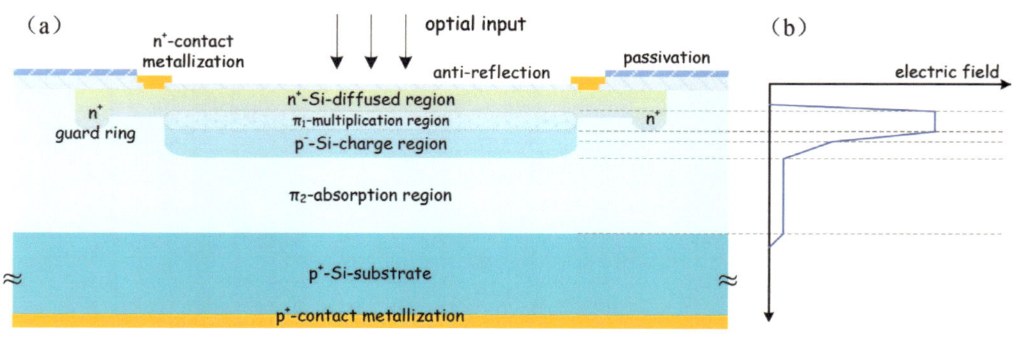

Figure 2. (a) Schematic cross section (not to scale), and (b) schematic profile of electric field of a typical APD structure.

of Si APDs on 6-inch or 8-inch wafers, as is now usually the case, will often mean that the absorption region is thick (~700 μm), operating voltages are high, and response times are slow. These problems can be avoided with the use of an epitaxial version of the design in **Figure 2(a)**. In this approach, a high-resistivity π-type layer is epitaxially grown on the surface of a low resistivity p-type Si substrate. The absorption region (epitaxial layer) may be of any thickness (typically chosen to be in the range of 30–50 μm), and its resistivity is chosen to be low enough so that it does not introduce a significant series resistance. When bias voltage is applied, the depletion layer stops at the interface between the substrate and the epitaxial layer [31–34]. While fast response is a requirement, the narrow active region of this APD is normally the best option.

Linear-mode Si APD arrays are fabricated by adopting Si planar manufacturing process on a high-resistivity π-type layer, which is grown epitaxially on the top of a low resistivity p-type Si substrate. The initial material developed is the Si layer 35–40 μm thick, a highly-resistive epitaxial layer on a p^+-type Si <111> substrate.

2.2. Design of the TIA

As mentioned above, the ROIC chips consist of a voltage-stabilized source and a TIA. The voltage-stabilized source effectively reduces external noise jamming and increases voltage suppression ratio of the power source. Here we have the structure of a new-type regulated cascode circuit configuration which is compatible with the APD chips. The bandwidth, the parallel negative feedback and the trans-impedance gain of the TIA are improved by using regulated cascode circuit [35, 36].

Figure 3 schematically illustrates the TIA with the regulated cascode circuit configuration. The regulated cascode circuit consists of common gate amplifier input stage (including R_1, R_2

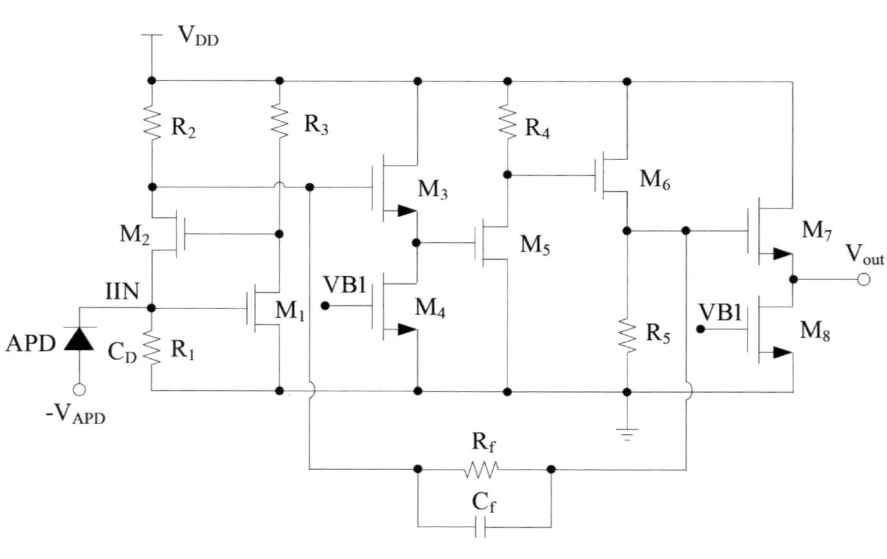

Figure 3. The TIA with the regulated cascode circuit configuration.

and M_2) and common source amplifier stage (including M_1 and R_3), forming partial current parallel negative feedback. The primary photocurrent generated by APD is imported to the source electrode of the cathode-input amplifier. Then the APD's current signal is converted into a voltage signal. The source follower, consisting of M_3 and M_4, is used for isolation. The secondary common-source amplifier is composed of M_3 and R_4, which play a part role in amplifying signal again. To improve the output drive capability, the output stage contains two-stage source followers, made up of $M_6 + R_5$ and $M_7 + M_8$ respectively.

2.3. Properties of the Si APD array

Developed at SITP, Si APD arrays were characterized at UESTC. The fabricated devices exhibit high primary photoelectric sensitivity (about 0.5 A/W @905 nm at gain M = 1) and high speed of operation (about 10 ns). **Figure 4** shows an example of typical dependences of the gain on the reverse bias. As the bias arises up to the reach-through voltage V_{rt}, it depletes the π-type avalanche region. For the APDs, V_{rt} attains values of 60–70 V, over which not much more regions are depleted. Further increasing the bias voltage mainly leads to higher electric fields in the structure. As the highest electric field reaches the critical value E_{cr}, multiplication of carriers starts to occur. More rising in the reverse voltage makes the steady current density go up to in principle infinity, where actually the avalanche breakdown takes place [37–39]. The corresponding voltage here is thus named avalanche breakdown voltage V_{br}, about 110 V for Si APDs.

At operating voltage (V_{br} × 98%), the multiplication region of a Si APD has an electric field as high as about 3.7×10^5 V/cm and an impact generation rate as high as about 2.8×10^{25} s^{-1} cm^{-3}. As a result, the avalanche gain (*M*) and the sensibility (*S*) of the linear-mode Si APDs are observed to be up to about 600 and 300 A/W @905 nm respectively.

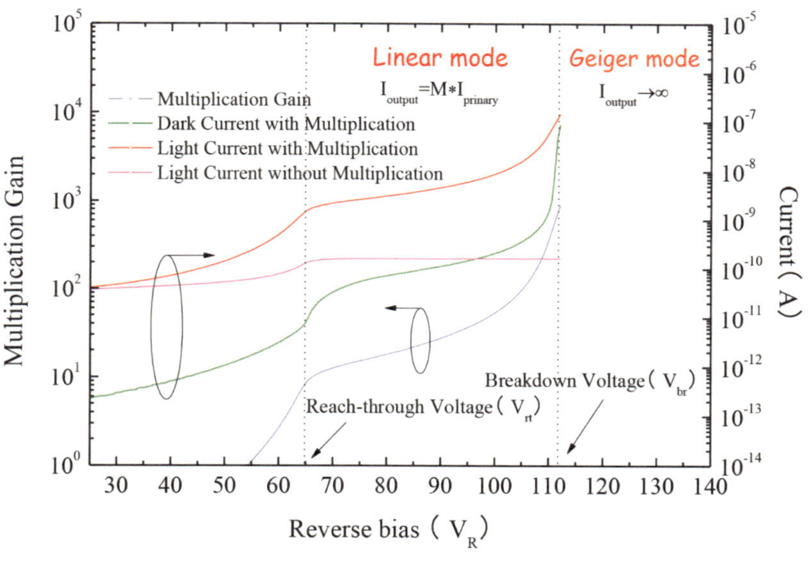

Figure 4. Dependence of dark current, photocurrent (with/without multiplication) and multiplication gain on reverse bias voltage of the designed SACM Si APD.

The ROIC chips were developed on the 0.18-μm CMOS platform of SMIC, Shanghai. The voltage-stabilized source effectively reduces external noise jamming and increases voltage suppression ratio of the power source. TIA shows trans-impedance of 120 dBΩ, the equivalent input noise is about 6 pA/Hz$^{1/2}$, the rise time is 7.3 ns, and the bandwidth is $BW \geq 35$ MHz.

Arrays of 64 × 1 Si APDs and ROIC chips were integrated to form the photodetector device by performing bonder-leading welding techniques. Together with packaging processing, the devices of 64 × 1 Si APD focal-plane arrays were successfully fabricated, one of which is shown in **Figure 5**. The power of input signal light is 0.9 nW (the duty cycle is 1/1000), and the maximum output voltage amplitude is 1.04 V. The devices present pulse responsivity $R \geq 1 \times 10^6$ V/W, noise equivalent power $NEP \leq 5$ pW/Hz$^{1/2}$, rise time $t_r \leq 3$ ns, and inhomogeneity of responsivity of each pixel ≤10%, under 905 nm, 100 ns and 10 kHz of laser irradiation.

2.4. Application of the linear mode Si APD array

As we constructed linear-mode Si APD focal-plane detectors, the 64 × 1 array devices are tested for possible applications. One example is that, the device is effective in running a driverless platform. Using this APD array, an obstacle-avoidance LADAR, as shown in **Figure 6(a)**, is successful with detection distance of 110 m, distance resolution of 5 cm and angle resolution of 0.5°. This LADAR can effectively detect the obstacles on the way, as shown in **Figure 6(b)**. Compared with traditional technique, in which a single detector was used, the image is much clearer (10 times of pixels) and the imaging speed is much faster (35 versus 15 Hz), so this newly developed obstacle-avoidance LADAR is more accurate and better to be used in driverless vehicles.

Figure 5. The device of 64 × 1 linear-mode Si APD focal-plane array.

http://dx.doi.org/10.5772/intechopen.81294

(a)

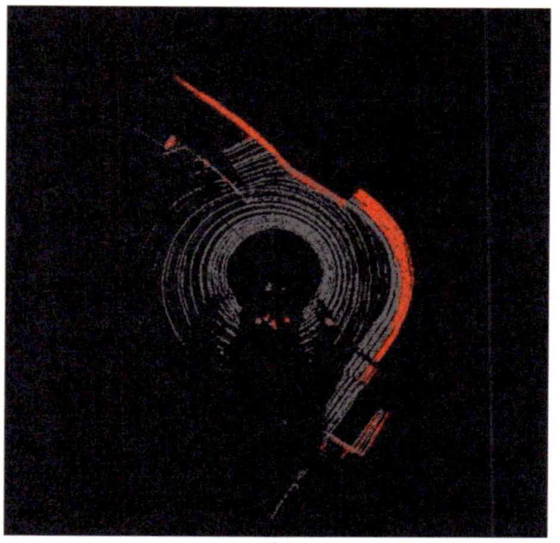

(b)

Figure 6. (a) An obstacle-avoidance LADAR, using 64 × 1 linear-mode Si APD array as the focal-plane detector, is installed on a driverless vehicle. (b) The imaging effect of the LADAR, where the red pattern shows the existence of obstacles.

3. Si SPAD focal-plane arrays

A Geiger-mode APD can detect a signal as weak as a single photon. In recent years, it is very active and effective as a single-photon detector and usually termed SPAD. Organized into arrays, SPAD can be used in many systems such as LADAR, mobile laser imaging and viewing instrument. By using some special processing, we developed typical Si-based SPAD arrays working at 905 nm. The key techniques are described as follows.

3.1. Fabrication of Si SPAD array chip

3.1.1. Design

According to the requirement of a Si SPAD, there would be a depletion region as thick as 30 μm. Using usual single-sided abrupt p-n junction, to get such a large depletion while remaining avalanche gain, one need to apply a voltage as high as 500 V, which is not realistic enough. We design a reach-through structure, containing n^+-π-p-π-p^+ layers, as shown in **Figure 7**. The electric field distribution under bias near breakdown is similar to that in **Figure 2(b)**. The light-generated carriers is multiplied in the region with highest electric field, so called multiplication region. This region is very thin compared to the whole depletion region. The other parts in the depletion region can have electric field as weak as possible but sufficient to ensure carrier drifting at the saturated speed. As a result, the operating voltage can be greatly decreased.

There could be two types of host materials. One is high-resistive Si wafer (p-type with 10^{14} cm^{-3} boron doped), the other is epitaxial lowly-doped p-type Si on a p^+-doped Si wafer. To fabricate the reach-through structure on bulk Si, the whole wafer has to be depleted as the

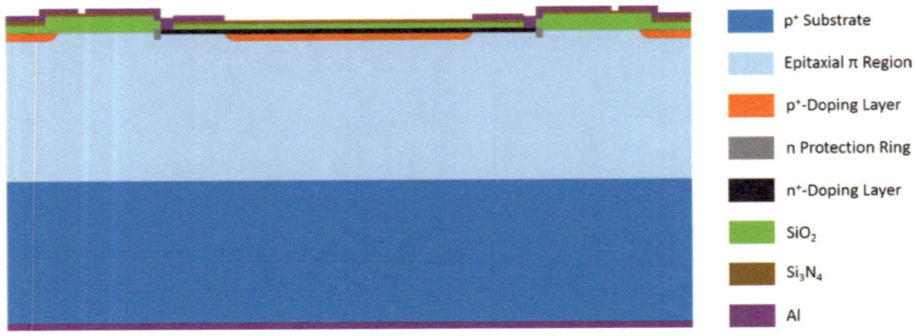

Figure 7. The designed structure of Si SPAD.

device works. To get low breakdown voltage of 150 V or so, we must polish the wafer down to a thickness of 50 μm, which is difficult and brings a lot of unstable features to the devices. Therefore, we use epitaxial wafer as the material for fabricating Si SPAD arrays.

Decreasing the dark count rate (DCR) is conflicting with choosing lower avalanche electric field E_{max} and lower operating voltage. There should be tradeoff between these two to get optimized structure parameters. Our previous product has DCR of several 10 kHz with light receiving area of Φ500 μm. Here we try making a device area of Φ50 μm to get much lower DCR. By further well designing and optimizing the internal structure, improving the pixel uniformity and surface passivation effect, we succeeded in controlling the DCR under 10 kHz.

3.1.2. *Precise control and uniformity*

Precise control of the device structure is a decisive step. The multiplication layer and charge layer are most important because they greatly influence the key parameters such as quantum efficiency, response time and gain. To precisely control the charge quantity and multiplication length, we adopt the following process. Boron ion-implantation is firstly performed to accurately control the dopant amount. Due to smaller diffusion coefficient of boron in Si, the broadening of boron distribution after thermal treatment is weaker. Then, second epitaxy is carried out to grow n-type layer. This is the way to precisely control the multiplication length.

It is the critical process of a Si SPAD array to make all the pixels controllably consistent in characteristics, e.g. the avalanche gain, the response time, and the breakdown behavior. The pixels uniformity is influenced by four factors: epitaxial structure, ion-implantation of the charge layer, diffusion of the p-n junction, heat-induced doping redistribution in the device process. The most important is that, the epitaxial layers should be grown as uniform as possible. Usually, a 50-μm thick epitaxial layer should have thickness uncertainty of less than 50 nm.

In an SPAD array, one of the critical structures is the guard-ring. The designed SPAD structure shows that, the n^+ contact and light-incident layer is so small that the p-n junction depth is about 0.5 μm. In order to prevent the device from being lowly broken down, a guard-ring around the device can be fabricated by doping at the edge of the n^+-doped area. It uses more deeply diffused n-doping (Phosphorus has big diffusion coefficient in silicon) to decrease the curvature rate of the edge of the p-n junction with the π-region, to reduce the electric-field of the junction edge, to improve the breakdown voltage at the edge of the device. Properly controlling diffusion depth, doping level and ring width, the edge breakdown voltage can be improved to be about 1.5 times that in the avalanche region. As an example, usually, 2–3 μm wide guard-ring is suitable for the above purpose.

3.1.3. Isolation process

In the avalanche procedure, there come photons at wavelength shorter than 1 μm while creating multiplication carriers. These photons may enter other pixels nearby to generate unintentional count. In order to suppress cross-talk between pixels, design of the SPAD structure is optimized. Further by processing the light-hiding belt and with the aid of decreasing the reflectivity at the interface, the optical cross-talk is well controlled. In the Si SPAD array, there exist big shunt capacitances in between the adjacent pixels, so the electric cross-talk could be of high possibility. This is resolved by decreasing capacitances related to the wires in the later interconnection process.

With a long absorption region (~30 μm), it needs to keep background doping level being lower than 10^{14} cm^{-3} in order to remain lower electric field in this layer. In the meantime, the inversion layer near the surface is a major factor causing device failure. Thus, it is necessary to set a p^{+}-doped area around the surficial active region, i.e. the channel-resisting region. It can cut the inversion layer at the SiO_2-p-Si interface and stop the surficial expansion of the depletion region. Its final role is to suppress the surficial leakage current and to prevent the device from being broken down by a low bias. By calculation and experience, a doping level higher than 10^{16} cm^{-3} in the surface layer is sufficient to form the channel-resisting effect.

By performing the above processes, chips of 32 × 32–64 × 64 Si SPAD arrays are fabricated. One sample is shown in **Figure 8**.

Figure 8. One chip of Si SPAD array.

3.2. ROIC optimization

To realize highly accurate timing of the photon arrival, we use a time-digit-conversion (TDC) circuit with the aid of phase-shifting technique. This approach satisfies the requirements of 2 ns in the time resolution and 20 kHz in frame frequency while decreasing the power consumption of the whole chip. An active-quenching design is used to reach an extinction time of less than 50 ns.

High precision timing and time-reading circuit is composed of TDC and memory readout unit, as shown in **Figure 9**. After gate/signal conversion circuit transforms the high voltage output of SPAD into low voltage signal, it firstly needs to read the photon flying time of every pixel through TDC, to change times into digits, and to read out the digits through the memory readout circuit. For the purpose of 2 ns of time resolution, highly frequent, highly stable main clock is supplied to the 12-digits counter. Frequency more than 500 MHz is not easy to be realized at every pixel at the same time. Due to the processing limit, shunt resistors and capacitors may significantly contribute to the power consumption (To an 32 × 32 array, the power consumed in clock lines would be 100 mW). Without using phase-locking loop (PLL), here we design TDC circuit with the aid of phase-shifting technique, to satisfy the requirement of high time resolution while decreasing the requirement for clock frequency. The TDC consists of counter and time-delay chain. When the external timing signal (a rising edge) comes, the counter starts to work; when a photon is detected, the circuit generates a signal to stop the counter and remain the present count data. Via the time-delay chain composed of 8 units, the external clock creates 8 clock signals with different phases. As the starting signal comes, every clock signal is sampled and the time-delay chain outputs an 8-digit signal, which will be coded and saved into the data process module. A similar process happens when a stop signal comes. Difference calculation between the start and stop data gives a 4-digit signal, constructing a 12-digit information together with the data from the counter. Then, the time interval

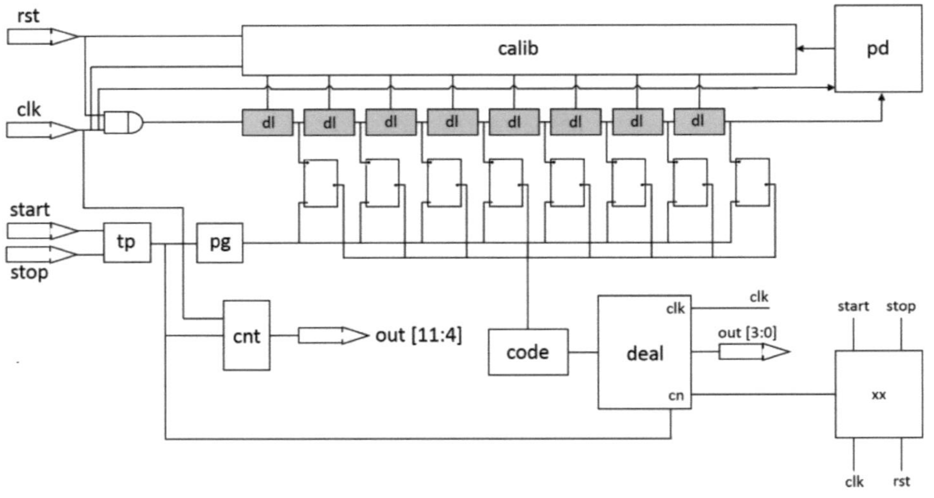

Figure 9. Structure of the high precision time-digit conversion circuit designed for Si SPAD array.

http://dx.doi.org/10.5772/intechopen.81294

between start and stop is converted into a 12-bit digital signal. When the counter output clock is effective, its locked state is transferred bit by bit into a 12-digit register. Controlled by a 25 MHz clock, the register transfers its digits into the register of the neighboring pixel. This serial transform mode gives at last a frame frequency of 25 kHz.

Avalanche-quenching circuit is another important aspect in SPAD array. To realize a dead-time less than 50 ns, we design actively-quenching circuit as shown in **Figure 10**. When avalanche photocurrent is detected, the voltage at point a jumps down and forces the quenching circuit run. After a while, the voltage at point b comes higher, switching on the transistor M1 via the feedback branch, and quickly pulling down the voltage at point V_{apd} to make the APD bias lower than breakdown voltage (quenching the APD). After more a while, the charge–discharge circuit gradually decreases the voltage at point b, M1 is turned off via feedback circuit, and V_{dd} charges the APD through R_0 and parasitic capacitances to restore the Geiger mode.

The above designs are realized by performing CMOS processing, and thus the ROIC chips are produced.

3.3. Interconnection and package

The next key processing is interconnecting the SPAD chip and the ROIC chip. The fabrication is as follows. After some degree of thinning processing, the backside of the wafer is treated to have a light-incident window for every pixel. As shown in **Figure 11**, the window area is formed by etching off the p-type substrate, and the layer under the etched window is made to be ~35 μm thick. Dry etching such as $SF_6 + O_2$ ICP is used to fabricate windows with straight sidewalls. In addition, the uniformity of this processing must remain ≤ ±2% in the window depth and ≤ ±0.5 μm in the window diameter. Then, it is realized that negligible light is absorbed by the p-type substrate. Using standard Indium-shot interconnection processing, the SPAD chip is connected exactly with the ROIC, as shown in **Figure 11**. Integrating the interconnected chip with TEC cooling cells, and packing these all into a vacuum can, a Si SPAD focal-plane device is developed.

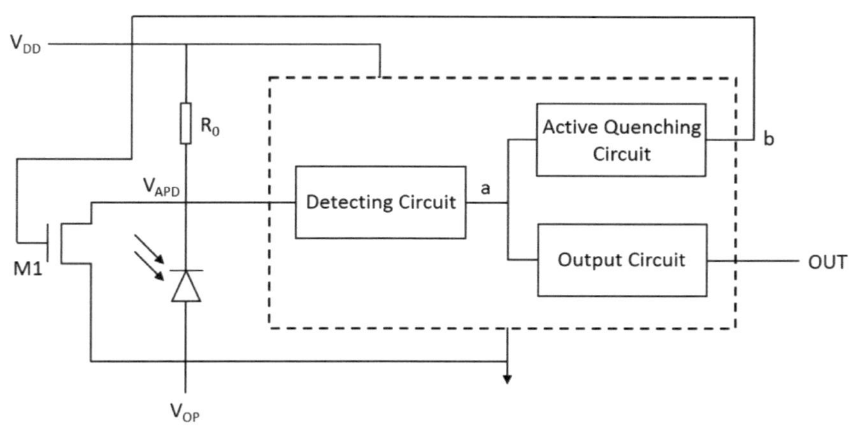

Figure 10. Structure of the quenching circuit for Si SPAD array.

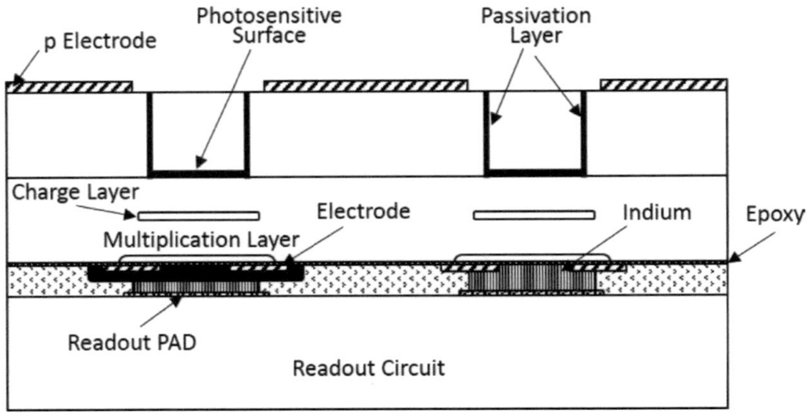

Figure 11. Schematic of the interconnection process for fabricating Si SPAD arrays.

3.4. Application of Si SPAD focal-plane arrays

Measurements show that the developed SPAD array have DCR lower than 5 kHz, average photon detection efficiency (PDE) ~25%, time resolution <1 ns, and frame speed ~25 kHz. The 32 × 32 SPAD array exhibits pixel uniformity < ±5% (e.g. in counting rate). It can thus be applied in a practical imaging system. The above fabricated SPAD array device is installed in a LADAR 3D imaging instrument. The instrument can three-dimensionally detect and recognize multi-hided objects in forests and mountains, and be of small size, light weight, high resolution and rapid imaging.

4. InGaAsP/InP SPAD focal-plane arrays

Like InGaAs/InP SPADs [40], InGaAsP/InP SPADs are also extensively studied and practically explored for their utility in many fields including single photon imaging [41] and quantum information processing [42, 43] in the near-infrared wavelength range. Thanks to the advanced epitaxial techniques, this kind of SPADs has been well developed and applied in, e.g., LADAR in recent years [41, 44]. Nevertheless, many critical problems are still open to be resolved. One of them is the device homogeneity, such as the reproducibility and uniformity of the SPAD performance [45], which are strongly required to be precisely controlled by refining the structure parameters in epitaxial growing and device processing. One can, of course, estimate the effects of some individual physical parameters on the performance homogeneity using some analytical method [46], but it is not easy to obtain the knowledge of many parameters at the same time. It is even unlikely to make clear the collective influence of multiple parameters and to quantitatively take a balance between various parameters. The quantitative association between the device inhomogeneity and structure uncertainty should thus be necessarily established. Therefore, we firstly carry out a statistical analysis on InGaAsP/InP SPAD characteristics by randomizing the structure parameters, and then figure out the necessary control accuracies in a few significant structure parameters, which are required for nice device homogeneity. Accordingly, we fabricate InGaAsP/InP SPAD focal-plane arrays and demonstrate their applications.

4.1. SPAD array homogeneity versus material controllability

An InGaAsP/InP APD structure is designed as an example SPAD object. It is of a hetero-structure comprising SACM layers, as can be seen schematically in **Figure 12(a)**. By using conventional APD theory [28, 47] and lately advanced approaches [48–50], citing material parameters from previous reports [49, 51, 52], and neglecting the dead-space effect [53], the device characteristics of this structure was calculated.

Figure 12(b) illustrates the calculated current–voltage (*I-V*) characteristics in dark. Here the breakdown voltage V_b is principally the applied reverse bias V at infinite avalanche current. The simulated DCR r_d versus PDE η is shown in **Figure 12(c)**. Both of them are dependent on the applied excess bias $V_{ex} = V - V_b$. As the SPAD is running under a middle excess bias $V_{ex0} = 5$ V, r_d is found below 10 kHz and η appears some 0.50. It suggests that 5 V of V_{ex} is an optimal operating condition at 230 K, so $V_{ex} = 5$ V will be the reference point in this study.

As a SPAD array is used, there is usually a common bias V_0 generally applied to more than thousands of pixels [44]. Provided structure fluctuations exist among the numerous pixels, the

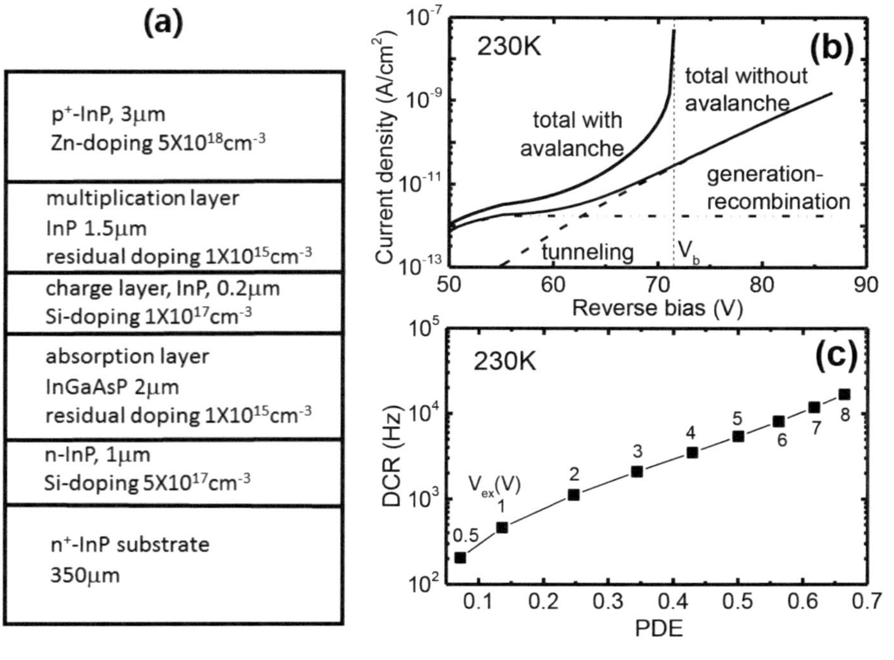

Figure 12. (a) The structure diagram of a designed InGaAsP/InP Geiger-mode APD (or SPAD) studied in this work; (b) calculated *I-V* relationship in dark and (c) calculated DCR versus PDE for the InGaAsP/InP SPAD device at 230 K.

effective V_{ex} will vary from this to that pixel so that device performance exhibits inhomogeneity. To clarify this effect, we first set any structure parameter t randomly changing in a way as

$$t_i = t_0(1 + W\sigma_i), \tag{1}$$

where the subscription $i = 1, 2, 3$ is the calculation sequence number in a series of simulations, t_i is the ith value of parameter t, t_0 is the designed value of t, W is the fluctuation degree of t relative to t_0, and σ_i is the ith value of the random variable σ, distributed in a normal mode with full width at half maximum (FWHM) of unity. Similarly we can set

$$\begin{aligned} n_{ci} &= n_{c0}(1 + W_{nc}\,\sigma_{nci}), \\ t_{mi} &= t_{m0}(1 + W_{tm}\,\sigma_{tmi}), \end{aligned} \tag{2}$$

and so forth, where $n_{ci}(t_{mi})$ is the ith value of charge density n_c (multiplication width t_m), $n_{c0}(t_{m0})$ is the designed value of $n_c(t_m)$, $W_{nc}(W_{tm})$ is the relative FWHM of $n_c(t_m)$ with respect to $n_{c0}(t_{m0})$, and $\sigma_{nci}(\sigma_{tmi})$ is the ith value of σ for changing $n_c(t_m)$. All the variables are defined in a similar way to the above. The structure parameters are changing independently because each has its own FWHM and σ values. With a set of structure parameters (n_{ci}, t_{mi},...), one set of device performance data is calculated. With thousand sets of device performance data, distributions of V_b, V_{ex}, r_d, and η are obtained through statistics, and then the correlation between the performance fluctuations and device structures is figured out.

Our simulations show that n_c, t_m and t_c (charge layer thickness) have strong effects, while absorption layer doping level n_a, thickness t_a and multiplication layer doping level n_m have weak effects on V_{ex}. The strong t_m effect can be easily understood since the width of the multiplication region is crucial to determine the characteristics of a SPAD [48, 54]. It means that the charge quantity should be controlled most precisely in design and epitaxy process. In addition, we see that every single structure parameter leads to V_{ex} fluctuation in a linear manner.

Extending the above simulation to more structure parameters, the homogeneity of device performance can be obtained in terms of varying parameter numbers. The result of V_{ex} is displayed in **Figure 13(a)**, where a sublinear change of V_{ex} fluctuation is seen to happen with increasing parameter number. Taking all of the six parameters into account, we get that V_{ex} varies with a FWHM of 24%, far less than a simple summation of the effects of individual parameters. The V_{ex} distribution arisen by six independently varying parameters is demonstrated in **Figure 13(b)**. With the referred excess bias $V_{ex0} = 5$ V, the practical value of V_{ex} varies mainly in the range of 4.4–5.6 V, quite good for many applications. It may be also worthy to get the effects on other performance characters. **Figure 13(c)** presents the variation of DCR r_d, which exhibits a roughly normal distribution with a wide relative FWHM (54%). In detail, the worst DCR is some 30% higher than the designed value, which is acceptable in applications. **Figure 13(d)** displays the distribution of PDE η, normal with a narrow relative width (17%). It suggests a PDE change within 0.46–0.54, which is homogeneous enough in many applications.

The reason why the DCR fluctuation is much larger than PDE is that DCR depends almost exponentially on V_{ex} and PDE, as **Figure 12(c)** shows.

In realistic epitaxial growth, controlling the thickness or the doping level may usually have a common precision in different layers, although a non-dope layer would have worse carrier density fluctuation than the intentionally doped regions. The following thus shows a way more effective to investigate the uncertainty correlation between the epitaxy growth and the device performance. Let us study the device characteristics varying with common fluctuation width W_{nd} of the residual carrier densities in the absorption layer and the

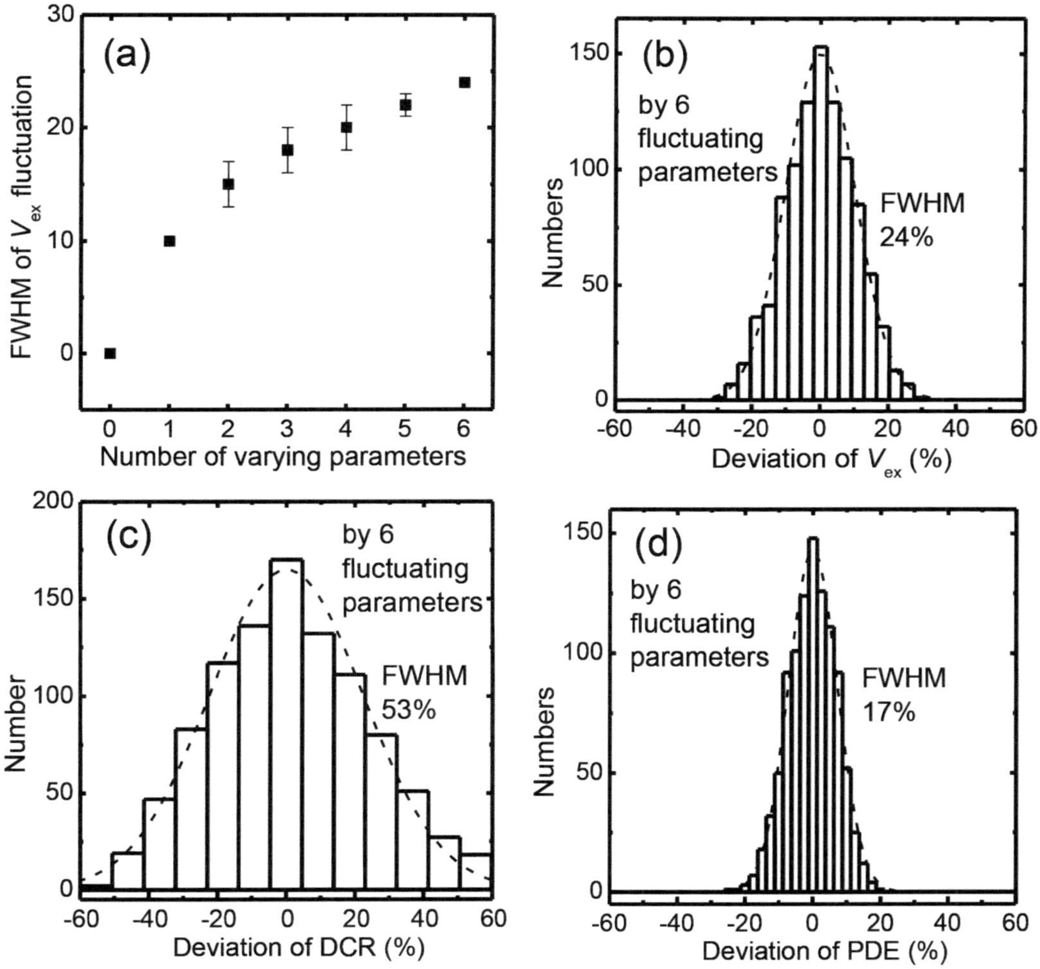

Figure 13. (a) Simulated distribution FWHM of the excess bias V_{ex} dependent of the number of simultaneously varying parameters; distribution histograms of (b) V_{ex}, (c) DCR and (d) PDE brought about by six simultaneously varying parameters including the doping level and the width of the absorption region n_a and t_a, of the charge layer n_c and t_c, and of the multiplication region n_m and t_m, respectively. The dashed lines on the histograms indicate the fitted results to normal distributions.

multiplication layer, common fluctuation width W_t of the widths of the absorption, charge and multiplication regions, and fluctuation width W_n of the doping level in the charge layer, that reads

$$\begin{aligned} n_{ai} &= n_{a0}(1 + W_{nd}\,\sigma_{nai}),\\ n_{mi} &= n_{m0}(1 + W_{nd}\,\sigma_{nmi}),\\ n_{ci} &= n_{c0}(1 + W_n\,\sigma_{nci}),\\ t_{ai} &= t_{a0}(1 + W_t\,\sigma_{tai}),\\ t_{mi} &= t_{m0}(1 + W_t\,\sigma_{tmi}),\\ t_{ci} &= t_{c0}(1 + W_t\,\sigma_{tci}). \end{aligned} \tag{3}$$

One typical trial is to examine the dependence on two fluctuating parameters while remaining the others fixed. With W_{nd} fixed, **Figure 14** shows the V_{ex} contours as functions of W_n and W_t. From these data, the precision range of epitaxy growth required for definite performance homogeneity can be clearly seen. At first, two conditions with 10 and 20% of W_{nd} appear close to each other, especially for high V_{ex} variations, owing to the weak effect of the carrier density in non-dope layer. The relationship between W_n and W_t is far away from a linear curve but more like a circle for a finite V_{ex} fluctuation. To constraint V_{ex} fluctuation below a certain value, the fluctuations in thickness and charge control should roughly follow

$$W_n^2 + W_t^2 < W_x^2, \tag{4}$$

where W_x is a certain precision control value of thickness and charge. Quantitatively speaking, V_{ex} relative fluctuations below 50, 40, 30 and 20% need W_x values of about 4, 3, 2 and 1%, respectively.

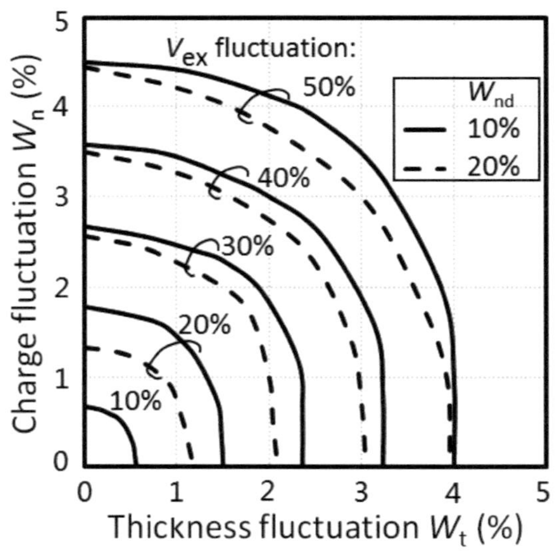

Figure 14. Contours of V_{ex} fluctuation width as a function of the thickness and doping level fluctuation widths W_t and W_n, under different fixed fluctuation widths of the residual carrier density in the non-dope layers W_{nd}.

In conventional growth, it is more difficult to control doping than thickness. Based on the result of **Figure 14**, as the thickness control can be better than 1–2%, V_{ex} homogeneity of 50, 40 and 30% could be realized by constraining the charge precision within 4–4.5, 3–3.6 and 2–2.7%, respectively. Viewed from another angle, the result is suggestive of a large space to tradeoff between the controls in thickness and charge. The example of V_{ex} fluctuating below 50% with W_{nd} = 10% suggests that the thickness (charge) precision $W_t(W_n)$ is better to be as small as 1% if $W_n(W_t)$ just satisfies 4.5%(4%), while $W_t(W_n)$ could be roughened to 3%(3.5%) if $W_n(W_t)$ weakly decreases to be about 3.5%(3%). In general, limiting the device inhomogeneity (in term of V_{ex}) below 50, 40, 30, and 20% needs the thickness and charge be controlled to a precision degree better than 3–3.2, 2.4–2.6, 1.7–1.9, and 0.9–1.2%, respectively. Since these degrees of control accuracies are easy or possible in epitaxy growth, InGaAsP/InP SPAD arrays are now producible in many laboratories [55–57] including our group, as will be described below. In order to finely limiting the device homogeneity, such as with V_{ex} fluctuation less than 10%, the thickness and charge should be controlled better than 0.5% in fluctuation, together with non-dope carrier density controlling within 10%. This degree of epitaxy precision is quite a challenging technique. It is possibly one of the reasons why it is presently still difficult to prepare 512 × 512 or larger scale SPAD arrays. Obviously, the above method is very helpful and effective to quantitatively correlate the controllability of multiple structure parameters with the SPAD device homogeneity.

4.2. Fabrication of InGaAsP/InP SPAD arrays

InP based APDs must use epitaxial materials. Firstly, we prepared APD materials with the main structure as shown in **Figure 12(a)** by using metal organic chemical vapor deposition (MOCVD). MOCVD growth is performed to satisfy the material uniformity requirements described above. On this epitaxial wafer, SPAD device structure as shown by **Figure 15** will be fabricated. As an array, there are isolating grooves (channels) between the pixels, Indium shots on the front side for interconnection and micro-lenses on the backside for light collection.

The key processes to fabricate the InGaAsP/InP SPAD arrays are as follows. First the active p-n junction is formed by selective diffusion. The diffusion process includes, thermally

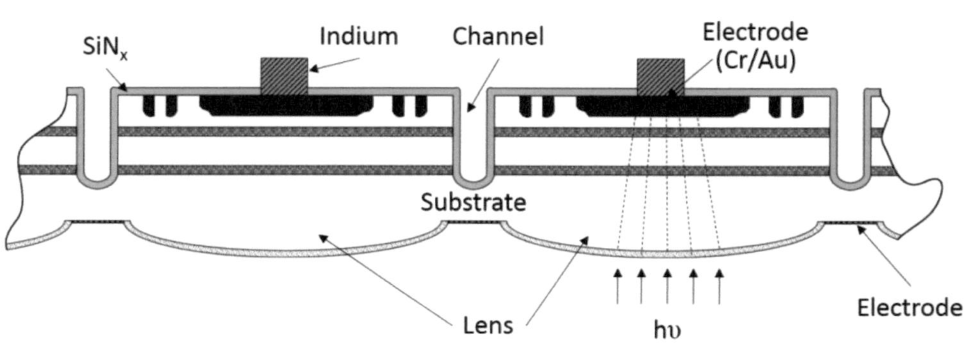

Figure 15. Structure of InGaAsP/InP SPAD array chip.

evaporating one layer of solid Zn_3P_2, depositing one layer of SiN_x to thoroughly cover Zn_3P_2, rapid thermal annealing to diffuse Zn into the chip, and etching off the SiN_x and the resident Zn_3P_2. To make the response time of a SPAD device shorter than 5 ns, the active area (the p-n junction area) of a pixel is made to be less than ϕ100 μm. Considering the guard-rings, the lateral depletion width and the diffusion length of electrons and holes, the distance between neighboring pixel centers is taken to be 300 μm. To suppress the cross-talk between pixels, the isolation between pixels is, besides the deep grooves, aided by highly resistive p-n junction. To increase the filling factor, light is incident on the backside, where there fabricated microlenses for each pixel. The microlenses here are not bonded onto the backside, but directly fabricated on the backside by specific dry-etching.

The fabricated SPAD array is characterized as shown in **Figure 16**. Measurements on material properties show that residual carrier density, layer thickness, and doping level fluctuations in a 10 × 10 mm^2 area appear about 8, 0.8 and 1.5%. On such a chip, 32 × 32–64 × 64 arrays of SPADs were developed and characterized at low temperatures. Under gated mode with gate repetition rate of 500 kHz and gate width of 10 ns, DCR and PDE were measured using a single-photon laser at 1.06 μm. The afterpulsing probability is controlled below 2% by setting the dead time to be ~2 μs. As presented by the inset in **Figure 16**, various pixels have dark *I-V* curves with V_b (defined to be the bias at 10 μA) weakly changing but around 75 V. **Figure 16** shows that, the fluctuation in the excess bias distributes in a normal way with FWHM of 14%, which is consistent with the simulated value 18%, and compatible with an estimation based on **Figure 16**. The DCR and PDE vary normally with FWHM of 31 and 10%, and this is also consistent with the simulated values 37 and 12%, respectively.

The ROIC is designed in a way similar to that of Si SPAD arrays. The interconnection is a standard indium-shot inversion-bonding process. After packaging into a vacuum can with

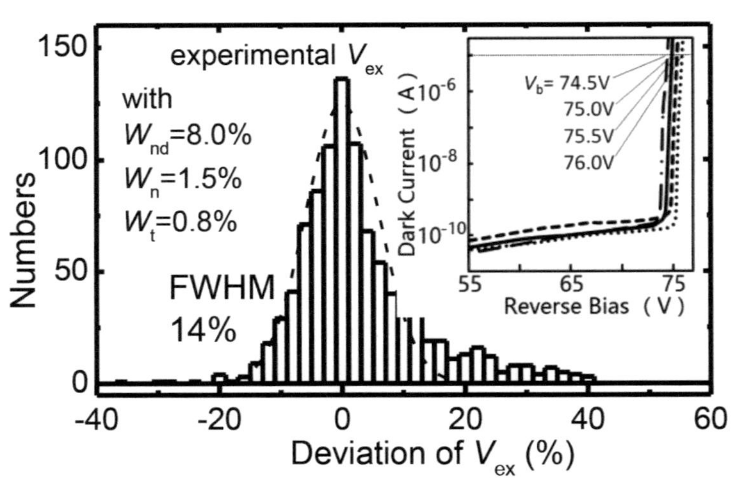

Figure 16. An experimental result of V_{ex} fluctuation distribution of InGaAsP/InP SPAD array. The dashed line represents the fitting to a normal distribution. The inset exhibits the *I-V* curves in dark of a few typical InGaAsP/InP SPAD devices.

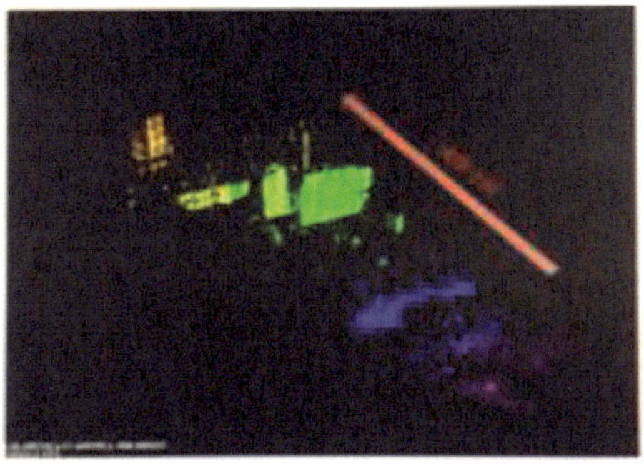

Figure 17. Imaging by a LADAR instrument with 64 × 64 InGaAsP/InP SPAD focal-plane array detectors.

a transparent window, the InGaAsP/InP SPAD array is developed and can be used in an imaging system.

4.3. Application of the InGaAsP/InP SPAD arrays

A 64 × 64 InGaAsP/InP SPAD array device is installed onto the focal plane of a LADAR system. Under 1.06 μm laser irradiation, the scence 1–3 km away was successfully imaged with 3D information, as shown in **Figure 17**.

5. Summary

APDs are being more and more widely and deeply studied to satisfy the requirement in weak light and single photon imaging. The progresses of this worldwide study, especially the distinctive researches and achievements in SITP and UESTC are reviewed. We successfully fabricated up to 64 × 1 linear-mode Si APD array, and 32 × 32–64 × 64 Si SPAD arrays, and applied them in LADAR platforms like driverless vehicles. Also, we developed 32 × 32-64 × 64 InGaAsP/InP SPAD arrays, and constructed 3D imaging LADAR using them. Together with the progresses of other groups and other materials, we see a prospective future for the development and application of focal-plane APDs.

Acknowledgements

This work was partially supported by 1000 Foreign Experts Program, China, and by the 1000 Talents Plan of Sichuan Province. We would like to thank Profs. L.B. Yu, Z. Shi, O. Wang, X. Li, Q. Dai, Y. Yang and Drs. J. Chen, X. Xie and Q. Xu at SITP for their practical help, and Profs. Zh. M. Wang, Q. Zhou and G.W. Deng at UESTC for their theoretical assistance.

Author details

Hai-Zhi Song[1,2]

Address all correspondence to: hzsong1296@163.com

1 Southwest Institute of Technical Physics (SITP), Chengdu, China

2 Institute of Fundamental and Frontier Sciences, University of Electronic Science and Technology of China (UESTC), Chengdu, China

References

[1] Campbell JC. Recent advances in avalanche photodiodes. Journal of Lightwave Technology. 2016;**43**(2):278-285

[2] Woodson ME, Ren M, Maddox SJ, Chen YJ, Bank SR, Campbell JC. Low-noise AlInAsSb avalanche photodiode. Applied Physics Letters. 2016;**108**:081102

[3] Campbell JC. Advances in photodetectors. In: Kaminow I, Li T, Wilner AE, editors. Optical Fiber Telecom-munications, Part A: Components and Subsystems. 5th ed. Vol. 5. San Francisco, CA, USA: Academic; 2008

[4] Bertone N, Clark WR. APD arrays—avalanche photodiode arrays provide versatility in ultra-sensitive applications. Laser Focus World. 2007;**43**(9):69-73

[5] Mitra P, Beck JD, Skokan MR, Robinson JE, Antoszewski J, Winchester KJ, et al. Adaptive focal plane array (AFPA) technologies for integrated infrared microsystems. In: Proceedings of SPIE 6232. SPIE - The International Society for Optics and Photonics. 2006. pp. 62320G-1-62320G-11

[6] Tosi A, Calandri N, Sanzaro M, Acerbi F. Low-noise, low-jitter, high detection efficiency InGaAs/InP single-photon avalanche diode. IEEE Journal of Selected Topics in Quantum Electronics. 2014;**20**(6):3803406

[7] Jiang XD, Itzler M, O'Donnell K, Entwistle M, Owens M, Slomkowski K, et al. InP-based single-photon detectors and geiger-mode APD arrays for quantum communications applications. IEEE Journal of Selected Topics in Quantum Electronics. 2015;**21**(3):3800112

[8] Aull B. Geiger-mode avalanche photodiode arrays integrated to all-digital CMOS circuits. Sensors. 2016;**16**:495. DOI: 10.3390/s16040495

[9] Othman MA, Taib SN, Husain MN, Napiah ZAFM. Reviews on avalanche photodiode for optical communication technology. ARPN Journal of Engineering and Applied Sciences. 2014;**9**(1):35-44

[10] Diagne M, McIntosh A, Donnelly J, Myszka M, Frechette J. Advances in InP/InGaAs Geiger-mode APD focal plane arrays. In: Proceedings of SPIE 10659. SPIE - The International Society for Optics and Photonics. 2018. p. 1065904

[11] Itzler M, Salzano G, Entwistle M, Jiang XD, Owens M, Piccione B, et al. Asynchronous Geiger-mode APD cameras with free-running InGaAsP pixels. In: Proceedings of SPIE 10212. SPIE - The International Society for Optics and Photonics. 2017. p. 102120K

[12] Cai X, Wu C, Lu H, Chen Y, Zhou D, Liu F, et al. Single photon counting spatial uniformity of 4H-SiC APD characterized by SNOM-based mapping system. IEEE Photonics Technology Letters. 2017;**29**(19):1603-1606

[13] Steindl B, Hofbauer M, Schneider-Hornstein K, Brandl P, Zimmermann H. Single-photon avalanche photodiode based fiber optic receiver for up to 200 Mb/s. IEEE Journal of Selected Topics in Quantum Electronics. 2018;**38013**:3801308. DOI: 10.1109/JSTQE.2017.2764682

[14] Ceccarelli F, Acconcia G, Labanca I, Gulinatti A, Ghioni M, Rech I. 152-dB dynamic range with a large-area custom-technology single-photon avalanche diode. IEEE Photonics Technology Letters. 2018;**30**(4):391-394. DOI: 10.1109/LPT.2018.2792781

[15] Chen J, Yang Y, Guo CD, Kong FL, Kou X, Wang O, et al. Linear-mode linear arrays 16 pixel silicon avalanche photodiodes with high gain and low noise readout. In: Proceedings of SPIE 10462. SPIE - The International Society for Optics and Photonics. 2017. p. 104625E

[16] Song HZ, Deng J, Dai Q, Shi Z, Yu LB. Quantitative correlation between fabrication precision and device homogeneity of single-photon avalanche diodes. IEEE Transactions on Electron Devices. 2016;**63**(12):4845-4851. DOI: 10.1109/TED.2016.2618222

[17] Dai M, Li X, Shi Z, Dai Q, Song HZ, Tang Z, et al. Gain and noise properties of multi-gain-stage superlattice InGaAs avalanche photodiode. Infrared and Laser Engineering. 2016;**45**(5):0520009

[18] Dai Q, Deng J, Shi Z, Yu LB, Song HZ. Multi-parameter fluctuation effects on InGaAsP/InP Geiger-mode avalanche photodiodes. Advances in Computer Science Research. 2015;**10**:965-969

[19] Chen J, Dai Q, Shen B, Deng J, Xu Q, Shi Z, et al. Multi-parameter study on the performance fluctuation of InGaAsP/InP single photon avalanche diodes. In: Proceedings of 2016 IEEE International Conference on Integrated Circuits and Microsystems (ICICM 2016). IEEE - The Institute of Electrical and Electronics Engineers. 2017. pp. 186-190

[20] Hao X, Chen J, Kong F, Ke Z, Wang O, Yu LB, et al. Design and photoelectric properties of silicon Geiger-mode APD array. In: ALAT 2017, Shenzhen, China: Laser Technology Press; May 20-21, 2017

[21] Williams GM, Huntington AS. Probabilistic analysis of linear mode vs Geiger mode APD FPAs for advanced LADAR enabled interceptors. In: Proceedings of SPIE 6220. SPIE - The International Society for Optics and Photonics. 2006. p. 622008

[22] Kishimoto S, Yonemura H, Adachi S, Shimazaki S, Ikeno M, Saito M, et al. A fast X-ray detector using silicon avalanche photodiodes of 64-pixel linear array. Journal of Physics: Conference Series. 2013;**425**:062007

[23] Sultana A, Kamrani E, Sawan M. CMOS silicon avalanche photodiodes for NIR light detection: a survey. Analog Integrated Circuits and Signal Processing. 2012;**70**:1-13

[24] Niclass C, Rochas A, Besse P, Charbon E. Design and characterization of a CMOS 3-D image sensor based on single photon avalanche diodes. IEEE Journal of Solid-State Circuits. 2005;**40**(9):1847-1854

[25] Brian FA. Silicon Geiger-mode avalanche photodiode arrays for photon-starved imaging. In: Proceedings of SPIE 9492. SPIE - The International Society for Optics and Photonics. 2015. p. 94920M

[26] Silvaco International Software. Atlas User's Manual [Z]. Santa Clara, CA, USA: Silvaco; 2012

[27] Tsang WT. Semiconductors and Semimetals. New York: Academic Press; 1985. pp. 247-328

[28] McIntyre RJ. On the avalanche initiation probability of avalanche diodes above the breakdown voltage. IEEE Transactions on Electron Devices. 1973;**ED-20**(7):637-641

[29] McIntyre RJ. Multiplication noise in uniform avalanche diodes. IEEE Transactions on Electron Devices. 1966;**ED-20**(7):637-641

[30] Thomas O, Yuan ZL, Shields AJ. Practical photon number detection with electric field-modulated silicon avalanche photodiodes. Nature Communications. 2012;**3**(644):1-5

[31] Thomas O, Yuan ZL, Dynes JF, Sharpe AW, Shields AJ. Efficient photon number detection with silicon avalanche photodiodes. Applied Physics Letters. 2010;**97**:031102

[32] Wang W, Feng Q, Wu W, Xie YT, Wang Z, Feng SJ. Analysis and simulation of process and performance of silicon avalanche photodiode. Infrared and Laser Engineering. 2014;**43**(1):140-144

[33] Wang W, Yan LS, Wang C, Du CY, Wang T, Wang GY, et al. Analysis of separate absorption charge multiplication Ge/Si-APD. Infrared and Laser Engineering. 2015;**44**(4):1349-1353

[34] Wegrzecka I, Wegrzecki M, Grynglas M, Bar J, Uszynski A, Grodecki R. Design and properties of silicon avalanche photodiodes. Opto-Electronics Review. 2004;**12**(1):95-104

[35] Mukherjee PP. Design, performance analysis and application of a modified improved MOS regulated cascode circuit. Procedia Technology. 2012;**4**:477-481

[36] Ben-Esmael M, Mathew M, Hart BL, Hayatleh K. Technique for increasing the output impedance of CMOS regulated cascode circuits. Journal of Circuits, Systems and Computers. 2016;**25**(10):1650118

[37] Wegrzecka I, Grynglas M, Wegrzecki M. Spectral dependence of the main parameters of silicon avalanche photodiodes. In: Proceedings of SPIE 4516. SPIE - The International Society for Optics and Photonics. 2001. pp. 187-193

[38] Kunar R, Chamberlain SG, Roulston DJ. Two-dimensional computer simulation of the breakdown characteristics of a multi-element avalanche photodiode array. IEEE Transactions on Electron Devices. 1984;**ED-31**(7):928-933

[39] Webb PP, Mcintyre RJ. Multi-element reach through avalanche photodiodes. IEEE Transactions on Electron Devices. 1984;**ED-31**(9):1206-1212

[40] Boso G, Korzh B, Lunghi T, Zbinden H. Low noise InGaAs/InP single-photon negative feedback avalanche diodes: Characterization and applications. In: Proceedings of SPIE 9492. SPIE - The International Society for Optics and Photonics. 2015. pp. 94920Q-1-9 4920Q-11

[41] Kutteruf MR, Lebow P. 1541 nm Gm APD LADAR system. In: Proceedings of SPIE 9080. SPIE - The International Society for Optics and Photonics. 2014. pp. 908006-1-908006-7

[42] Tosi A, Zappa F, Cova S. Single-photon detectors for practical quantum cryptography. In: Proceedings of SPIE 8542. SPIE - The International Society for Optics and Photonics. 2012. pp. 85421U-1-85421U-1-8

[43] Zhang J, Itzler MA, Zbinden H, Pan JW. Advances in InGaAs/InP single-photon detector systems for quantum communication. Light: Science & Applications. 2015;**4**:e286-1-e286-13

[44] Itzler MA, Krishnamachari U, Entwistle M, Jiang XD, Owens M, Slomkowski K. Dark count statistics in Geiger-mode avalanche photodiode cameras for 3-D imaging LADAR. IEEE Journal of Selected Topics in Quantum Electronics. 2014;**20**:3802111-1-3802111-11

[45] Itzler MA, Krishnamachari U, Chau Q, Jiang XD, Entwistle M, Owens M, et al. Statistical analysis of dark count rate in Geiger-mode APD FPAs. In: Proceedings of SPIE 9250. SPIE - The International Society for Optics and Photonics. 2014. pp. 925003-1-925003-12

[46] Savuskan V, Brouk I, Javitt M, Nemirovsky Y. An estimation of single photon avalanche diode (SPAD) photon detection efficiency (PDE) nonuniformity. IEEE Sensors Journal. 2013;**13**:1637-1640

[47] McIntyre RJ. Multiplication noise in uniform avalanche diodes. IEEE Transactions on Electron Devices. 1966;**13**:164-168

[48] Ramirez DA, Hayat MM, Itzler MA. Dependence of the performance of single photon avalanche diodes on the multiplication region width. IEEE Journal of Quantum Electronics. 2008;**44**(12):1188-1195

[49] Donnelly JP, Duerr EK, McIntosh KA, Dauler EA, Oakley DC, Groves SH, et al. Design considerations for 1.06-μm InGaAsP–InP Geiger-mode avalanche photodiodes. IEEE Journal of Quantum Electronics. 2006;**42**(8):797-809

[50] Javitt M, Savuskan V, Merhav T, Nemirovsky Y. Revisiting single photon avalanche diode current-voltage modeling and transient characteristics. Journal of Applied Physics. 2014;**115**:204503-1-204503-9

[51] Tan LJJ, Ng JS, Tan CH, David JPR. Avalanche noise of electrons and holes in (100)-oriented $Ga_{1-x}In_xAs_yP_{1-y}$. IEEE Journal of Quantum Electronics. 1985;**21**(9):1326-1337

[52] Itzler MA, Jiang XD, Ben-Michael R, Slomkowski K. InGaAsP avalanche photodetectors for non-gated 1.06 μm photon-counting receivers. In: Proceedings of SPIE 6572. SPIE - The International Society for Optics and Photonics. 2007. pp. 65720G-1-65720G-10

[53] Hayat MM, Saleh BAE, Teich MC. Effect of dead space on gain and noise of double-carrier-multiplication avalanche photodiodes. IEEE Transactions on Electron Devices. 1992;**39**(3):546-552

[54] Lee K, Yang K. Analysis of InGaAs/InP single-photon avalanche diodes with the multiplication width variation. IEEE Photonics Technology Letters. 2014;**26**(10):999-1002

[55] Bronzi D, Villa F, Tisa S, Tosi A, Zappa F, Durini D, et al. 100000 frames/s 64 × 32 single-photon detector array for 2-D imaging and 3-D ranging. IEEE Journal of Selected Topics in Quantum Electronics. 2014;**20**(6):3804310-1-3804310-10

[56] Smith GM, McIntosh KA, Donnelly JP, Funk JE, Mahoney LJ, Verghese S. Reliable InP-based Geiger-mode avalanche photodiode arrays. In: Proceedings of SPIE 7320. SPIE - The International Society for Optics and Photonics. Vol. 7320. 2009. pp. 73200R-1-73200R-10

[57] Itzler MA, Entwistle M, Owens M, Patel K, Jiang XD, Slomkowski K, et al. Comparison of 32 × 128 and 32 × 32 Geiger-mode APD FPAs for single photon 3D LADAR imaging. In: Proceedings of SPIE 8033. SPIE - The International Society for Optics and Photonics. 2011. pp. 80330G-1-80330G-10